UFOS DEJA VU
Skinwalkers, Shapeshifters, Alien Portals
And Vortexes To Other Dimensions

Conspiracy Journal
P R O D U C T I O N S

i

UFOS DEJA VU
Skinwalkers, Shapeshifters, Alien Portals
And Vortexes To Other Dimensions

By Timothy Beckley and Sean Casteel
With: Maria D'Andrea, Diane Tessman, Brent Raynes,
Tim Swartz, Tom Dongo, Linda Bradshaw, William Hamilton,
Erica Lukes, David Patridge, Nick Redfern, Chris O'Brien,
Dr. Leo Sprinkle, Joshua P. Warren, Albert S. Rosales
and Scott Corrales

Plus: Exclusive Interviews with George Knapp and Jeremy Corbell

Published in the United States of America By
Global Communications/Conspiracy Journal
Box 753 · New Brunswick, NJ 08903

Staff Members
Timothy G. Beckley, Publisher
Carol Ann Rodriguez, Assistant to the Publisher
Sean Casteel, General Associate Editor
Tim R. Swartz, Graphics and Editorial Consultant
William Kern, Editorial and Art Consultant

Sign Up On The Web For Our Free Weekly Newsletter
and Mail Order Version of Conspiracy Journal and Bizarre Bazaar
www.ConspiracyJournal.com

Order Hot Line: 1-732-602-3407
PayPal: MrUFO8@hotmail.com

CONTENTS

DEDICATION

This volume is dedicated to two hard working
individuals — Shutterbug, Charla Gene and
Adman Graphic Designer, William Kern.
Thanks for the assistance.

DEJA VU
OVER AND OVER

UFOS DEJA VU

**Publisher Timothy Green Beckley and Maria D'Andrea
at the book promotion in New York City.**

SECTION ONE
GATEWAY ENTRANCE

OUR SITE MAP

UFOS DEJA VU

INTRODUCTION

UFOS DEJA VU

INTRODUCTION

IT'S DEJA VU OVER AND OVER AGAIN
THE FEARLESS RAVINGS OF TIMOTHY GREEN BECKLEY
AND MARIA D' ANDREA

" It's deja vu all over again,!" was a famous Yogi-ism attributed to Yankee baseball hall of fame catcher Yogi Berra who explained that this expression originated when he witnessed Mickey Mantle and Roger Maris repeatedly hitting back-to-back home runs during the Yankees' seasons of the 1960's.

In less colorful lingo, deja vu means that one has lived through a situation before, whether it is in reference to walking down a street you "know" you have never walked down previously, but are doing it again, or sighting a UFO in pretty much the same local over and over and over as if it is wedged in the same spot in space. To see how deja vu might apply, I would like to reference Bill Murray in "Groundhog Day," and Tom Cruise who plays a soldier in a fierce battle against aliens who finds himself thrown into a time loop, in which he relives the same brutal fight – and his death – over and over, in a case of deja vu gone terribly wrong.

And while space does not permit us to reveal the location of the hundreds of spots where you might train your eyes on the sky and observe something very unusual hovering in the clouds, and perhaps lowering itself to the ground and interacting with you or your surroundings in various manners, both positively and negatively depending on the situation and your proximity to a particular portal, vortex or gateway. As far as I am concerned you may use the terms interchangeably, despite variances which we shall go into as we progress in our understanding of UFOs deja vu in general. In French the word means "seen before."

But let it be known that:

Some portals are permanent!

Other vortexes are temporary!

While some gateways come and go!

UFOS DEJA VU

I warn you that we are about to embark on a strange trip. It will be a lengthy voyage through hyperspace to other dimensions and parallel universes. There are some pretty weird — to our current trend of thinking — scenes, events and beings out there, wherever "there" might be referring to. But be fearless as we have a mighty crew at our book's helm to transport you to places you have never been before. As we go along they shall be identified. Most of these trekians I have ventured out into the galaxy with before.

One of the things that we have already predetermined fairly early is that we are not dealing with "ordinary" nuts and bolts craft from other planets. There is something decisively paranormal about UFOs that makes the topic much more complex and more difficult to prove a theory. Its advanced quantum physics for the general public. It circumvents normalcy and so you can call into play the supernatural, like it or not. As you will see by examining Tom Dongo's "19 Haunting Point Synopsis Of The Unknown" and Brent Raynes testimonial on UFOs and their relationship to the spirit world and life beyond death.

ENTER MARIA D'ANDREA

Over the years I have been known to consult our "in house" psychic and shaman Maria D' Andrea on the relationship between this world and all others. Maria is usually spot on with her thoughts and opinions when it comes to wormholes, portals and time slips, although she is not that connected with the UFO phenomena, "her thing" being more metaphysical.

When I told Maria "UFO deja vu" was in the works and that I was gathering thoughts and conjecture on portals, vortexes and stargates, she chimed right in and sent me her own thinking on the subject as set forth herein:

* * * * *

WHY DON'T WE JUST CALL THEM DOORWAYS?
By Maria D'Andrea MsD, D.D., DRH

Shamans and psychics use vortexes and portals all of the time. It's more or less how we get around. Going through them and exiting on the other side is our method for seeing what's going on within these other realms. They connect shamans and magi to other realms.

A vortex consists mainly of a flow with two main components: rapid velocity, mainly greatest at its center, and circulation. These in the paranormal field can be looked upon as carrying high positive psychic energy, as in some locations around the earth. These types of vortexes are mainly found on ley lines, or energy grids where a lot of UFO sightings and encounters take place.

When we are gathering energy to heighten our abilities which enable us to go dimension hopping, it's a method a lot of psychics have gravitated toward, passed down from generation to generation. I sometimes feel drained if I spend too much time near an active portal as we are not meant to be on such a high vibrational or

UFOS DEJA VU

alpha brain wave levels (psychic/intuitive) for too long a time. Limit your time there or all sorts of "crazy things" may begin to transpire, and you may feel anxieties and have to get yourself back to "normal" through tranquil meditation.

TRAVELING WITHOUT LUGGAGE

If you are visiting a portal or vortex you should take care to utilize psychic protection. Remember you are going to enter and interact with different realms. This is a doorway between our world and these parallel dimensions, distant places on our home planet and through time and space. Some could be hazardous to your health and well being.

Sometimes we visit portals of our own free will, other times they just open up without warning. And sometimes we can build them. On one of these occasions, I was out hiking by myself up in the mountains in Pennsylvania. As I walked through the woods, I heard a sound that seemed like a soft wind, and part of the landscape in front of me disappeared, and stars and planets became visible as if I were floating in space. As I looked, I saw what seemed to be two round disks flying erratically toward some destination. I "knew" there were beings inside these vessels and I wanted to see more. As I walked toward the portal, I had a feeling of being stopped as I got closer. It felt like I was trying to walk forward, but was only pushing into an etheric substance that felt like jello, and I wasn't going anywhere. It was very frustrating, because I'm a curious and active person as most sagas and shaman are. I automatically put up my psychic shield as a defense mechanism so that nothing could harm me. As I looked on, I noticed the "ships" were changing colors as they rotated. I didn't hear any sounds emanating from them. Nor did I see the beings inside, but I had an overwhelming feeling they were positive and meant no harm. I was lucky to have such a positive experience. Warning – not everyone does!

As I was gazing upon the setting before me and trying to gather more details about the ship and its occupants, the portal closed and I found myself back in the woods again.

To this day, I wonder what sort of beings were in the craft. I've opened doorways before and since, but couldn't get to the same place again. I think sometimes the universe just messes with you and they should have given me the proper coordinates. Right?

Do learn from your journey, because the vision quest is what this is all about. Remember, these vortexes and portals have been known and utilized in the spiritual fields for centuries, but may be new to the average UFO researcher. It feels good to know that science is just starting to delve into such matters as parallel dimensions, but we must learn the proper ways to deal with them so that we don't get burned. Some Ultra-terrestrials, as Tim calls them, mean us no harm; others can be quite negative, while others are just blundering into our realm without

UFOS DEJA VU

knowing what they are doing. There is, after all, a "trickster element" to much of this, as just about every Native America can tell you, if they are willing to discuss the subject.

* * * * *

Maria's concluding thoughts about Native Americans should be underscored, because a lot of the UFO deja vu phenomena takes place on or near sacred Indian ground throughout the Southwest, though it is not limited to this part of the country exclusively (or for that matter the rest of the world).

Now it is time to plot our course and set sail!

mrufo8@hotmail.com

ConspiracyJournal.com

Tom Cruise and Emily Blunt live and die over and over in the Sci-Fi thriller, "Edge of Tomorrow." >

Bill Murray relives the same day over and over until he learns the lesson in "Groundhog Day."

< Yankees Hall of Fame catcher Yogi Berra coined the phrase "Déjà vu all over again," when Micky Mantle and Roger Maris hit back to back home runs during their 1960 season.

SECTION ONE
PORTALS AND PASSAGES

UFOS DEJA VU

SECTION ONE

CHAPTER ONE
PORTALS, VORTICES, AND STARGATES
By Diane Tessman

CHAPTER TWO
VISITORS FROM THE DEAD ZONE
By Brent Raynes

UFOS DEJA VU

PORTALS, VORTICES, AND STARGATES
By Diane Tessman

PUBLISHER'S NOTE: With her silky long honey blonde hair and riveting green eyes, its no wonder, both earthmen and aliens alike are attracted to this Iowa based "first lady of UFOlogy" whose tenure in the field has been well established; initially as a MUFON investigator from Florida and later as abductee, channel and author. Her next literary offering is certain to turn a few heads and shake up those who think they have long ago discovered all there is to know about UFOs in our sky.

Diane Tessman's explosive new book, "Future Humans and the UFOs, Time for New Thinking," will be available on Amazon/Kindle in early 2020! Diane's new book revolutionizes the UFO field, offering new research and evidence, which have never been made public before! After 40 years in ufology, Diane Tessman offers her astounding conclusion – to the biggest puzzle of all! To which, she says, "The answer is meant to both shock and inspire!"

There are passageways between worlds, between life and death and between dimensions. Some are passageways in time which take us to the future or the past, while other corridors are shortcuts which will help us arrive at far distant planets thousands of light years away.

Each passageway, each corridor, has a door into which we must enter. Some of these passageways have doors at both ends which create a challenging barrier; we must have the bravery to pass through each of them - or have the wisdom to not pass through them.

There are portals, vortices (vortexes), and stargates. It is helpful to know the difference in definitions, even though we often use these words interchangeably.

* * * * *

PORTALS:

As a person dies, her spirit passes on through the portal which separates "physical life," from "spirit life."

The spirit life universe is much more exciting, diverse, peaceful and beautiful than the physical world. The portal to this world manifests as the soul leaves the physical body. Some emerging spirits perceive this portal as a brilliant light, while others simply sail toward colorful cosmic fractals, nebulas, and magnificent scenes beyond description.

There are other types of portals through which both angels and demons can travel from their own realms. These portals exist where you least expect them; just as the "physical life to spirit life portals," the "angel and demon portals" do not stay in a specific spot. Portals fluctuate, they manifest as needed by an entity. This entity can be a human spirit traveling from physical life to spirit life, or it can be on the request of an angel or other entity.

In quantum terms, portals are therefore created by the entity who needs them to be created. This ability in a dying human is part of the death (and back to universal new-life) process itself. It is a quantum power we all have, as do animals at that special time of physical death. And, beings in other dimensions of Earth have this power too; angels, demons, fairies, and other paranormal or other-worldly entities.

VORTEX: A vortex occurs when Nature's energies become supernatural energies and intersect with other natural or supernatural energies. Why do Nature's energies become super-nature? That is something only the mind of Nature knows. Why does a ghost light appear on a mountainside, or why does a will-of-the-wisp glide down the valley? Cosmic beings, even UFO occupants, have sensors which detect where/when a vortex appears, and they can make use of such a corridor.

A vortex usually stays in the same place, thus, as in the account of "Picnic at Hanging Rock," there is a specific area in rock formations which remains a "magic door" for years or at least for a while.

"Picnic at Hanging Rock" is an Australian historical fiction novel by Joan Lindsay, written 1967. It is based on a true event which happened around 1900, but the author has refused to explain if it is closely or loosely-based on a true event. Of course, Aboriginal Australia is full of magic and supernatural sites.

Set in 1900, "Hanging Rock" is about a group of female students at an Australian girls' boarding school who vanish at Hanging Rock while on a Valentine's Day picnic, and the effects the disappearances have on the school and local community.

Ominous-looking and mysterious rocks do exist at a real place called Hanging Rock in the Australian state of Victoria. These rocks may well contain a vortex to other dimensions, somewhere or some-when. Apparently in the real Hanging Rock occurrence, the girls and their governess who disappeared, were never seen again.

I investigated a similar rock formation in Joshua Tree National Park, in the High Desert of California. The local story had been circulating that a teenage couple disappeared there. I certainly felt the unusual energies of this rock formation, but

UFOS DEJA VU

I hesitantly stepped through the suspected vortex in the rock formation, and I am still here. Perhaps this vortex comes and goes.

Ley lines are invisible electromagnetic lines of Mother Earth and when Nature's energies create supernatural vortices, it is almost always because several ley lines have intersected at this spot. And so, here is a major difference between a portal (created by an entity's mind) and a vortex (created by nature, usually connected to ley lines).

STARGATE: The definition of a stargate is indeed different than a portal or a vortex. A stargate is space and/or time-connected whereas portals and vortices are usually Earth energies.

A stargate is often in the form of a wormhole but not all wormholes are successful stargates. Consider " Star Trek" for a moment: When the Enterprise goes to warp, it creates for itself a stargate through its advanced technology.

The starship folds space-time in this warp-corridor like folding a table cloth, and so the vast distances of space become relatively short distances. Stargates can also be used for traversing time, to the past or the future. There is currently much excitement around quantum "metamaterials" which physicists feel might be the key which UFOs use to fold space and time.

Wormholes are infrequent natural occurrences out in the universe, or they can be artificially created, as the Enterprise does. NASA works on creating artificial wormholes just as they seek to create warp drive; it is said it is easier to create a wormhole than it is to keep it open. If the corridor cannot be kept open, it is catastrophic for the crewmembers of the starship.

In "Star Trek: the Motion Picture," the Enterprise falls into a "wormhole" of imbalance when warp drive is engaged before it is ready. This suggests that falling into a wormhole is a failed attempt to go to warp; however, for NASA scientists, a wormhole is the method – the gate – to the stars and to time itself. Wormholes are stargates.

It is suspected that there are natural wormholes out in the galaxy but we have yet to find one, so for now, those natural "gates" are theoretical.

One day we will perfect the creation of wormholes into which starships can dive, to begin their journey to far distant worlds in a relative short time, as well as journeying though time itself.

Visit Diane at

www.EarthChangesPredictions.com

UFOS DEJA VU

Openings to a portal or vortex may be on the ground or in the sky at any altitude. Just recall how many planes vanished into the Bermuda Triangle, perhaps the biggest vortex of all.

Researcher/author/channel, Diane Tessman is fully conscious of the many portals and stargates existing around us.

Below: A still from the 1975 movie. Everyone at Appleyard College for Young Ladies agreed it was just right for a picnic at Hanging Rock. After lunch, a group of three of the girls climbed into the blaze of the afternoon sun, pressing on through the scrub into the shadows of Hanging Rock. Further, higher, till at last they disappeared.

Ask the crew of the Enterprise how they get from "here" to "there."

UFOS DEJA VU

VISITORS FROM THE DEAD ZONE
By Brent Raynes

PUBLISHER'S NOTE: UFO abductee, channeler and former MUFON investigator, Diane Tessman and I wrote a book several years ago "UFOs: Are They Your Passport to Heaven And Other Unearthly Realms?" In this study guide we detailed the reports of those claiming to have established some sort of contact or communication with their deceased loved ones during the course of a close encounter. Usually they would find themselves on board a UFO and there, in the company of some sort of alien being, would be a deceased loved one, a friend or usually a member of their immediate family. Some folks thought that we were disrespecting the dead, that we were being sensationalistic in claiming such a thing. That was not our intentions of course. We were merely passing along what we thought was valuable information we had been privileged to gather from what we were convinced were reliable sources, mainly those who had undergone such overly emotional experiences, making them burst out in tears.

When I interviewed Kenneth Arnold's granddaughter on our "Exploring the Bizarre" podcast, we discussed how Arnold, the man who is credited with coining the term "flying saucer," was convinced that UFOs were here removing the souls of the deceased to some other dimension (perhaps heaven?). This we tied in with the fact that over a dozen UFO investigators and authors had passed away on June 24, the anniversary of Ken's sighting, proving out a synchronicity that there is a tie in between UFOs and the dead. Zombies from outer space? B-movie film maker Ed Wood must have known what he was talking about when he produced, "Plan Nine From Outer Space," which has been a long running cult cinema classic. Tune into the interview with Shanelle Schanz, Arnold's grandaughter, on YouTube at – https://www.youtube.com/watch?v=lM__V_pX0UM&t=260s

We will be introducing Brent Raynes more thoroughly further on – he has a load of impressive credentials as you will see – but he did send us—"by accident"— a paper he published recently in his on line "Alternative Perceptions" newsletter, concerning this very crucial topic. We figure that early on would be a perfect place to establish the UFO Deja Vu and "dead zone" connection. It seems

UFOS DEJA VU

like the concept of UFO portals, we are about to expand upon may lead to what the spiritualists long ago identified as "the Summerland," or the place we go when we pass from this mortal coil.

x x x x x

Ann Strieber, the wife of "Communion" author Whitley Strieber, wrote an article in which she explained how her theory of "the visitors" is different from the majority in mainstream UFOlogy. In her article, she described how for years she and her husband had received large volumes of mail from people who have connected spirits of the dead with the UFO visitors. Back at the beginning, she began to make a list of commonalities that she would come across in the hundreds of letters that came pouring in after the publication of Whitley's best seller. One particularly struck her because she had not read it in the UFO literature was this: "They have something to do with what we call death."

After the publication of "Communion," Whitley got a phone call from his agent who had received a frantic message from a government official who said he urgently needed to talk to Streiber. Curiously – because of the source of the call – Whitley went ahead and called the man, who began by explaining, "I have to know what you think about this. I have to know if this is true." He then described how his wife stepped outside to walk their dog when a huge white light flew over their backyard and their little boy came running down the stairs to announce that his older brother, who had died a short time before in a tragic auto accident, had appeared to him "with two aliens, and he said to tell you he's all right. "

In another instance, a woman wrote to the Strieber's to describe how she and her young daughter were abducted and lost about two and a half hours. One night the daughter woke up screaming and then told the mother that the alien beings on the ship had informed her that she had—the aliens explained to her—"little bugs" that had entered her body and that she was going to die.

With that she slipped back off to sleep. The next morning, the child woke up with a high fever and severe swollen joints. It was discovered that she had a rare form of cancer called neuroblastoma, which involved the nervous system. A few months later she passed away. Just the day after her funeral, a friend called, and sounded frantic insisting on seeing the woman, and explaining that something had happened and she had to tell her about it.

When they got together they went out for pizza and the friend, feeling very awkward and like she was losing her mind, began to explain the urgent need for meeting with her. Something very peculiar had happened. Two nights before, around 2:20 in the morning, she described how two alien beings had appeared to her, and in between them was this woman's daughter. They wanted her to let this grieving mother know that her daughter was doing fine and was with them. After hearing this, the mother then shared with her bewildered friend the secret she

16

had told no one since it had happened, and explained about the alien abduction that she and her daughter had both experienced. Then the two friends both cried together.

While interviewing an Alabama ghost hunter for an issue of my "Alternative Perceptions" newsletter, this researcher remarked to me how he would not be surprised at all if it turned out that alien beings and ghosts used the same dimension to travel between this world and theirs. This merging of experiences and beliefs, that in the past were strictly kept within their respective and separate categories by those of various disciplines, seems to gradually be catching on with some. Maybe change is in the air.

Jump to "Exploring the Supernatural" episode on "Mr. UFOs Secret Files" on YouTube.com https://www.youtube.com/watch?v=1YvspDC9Jko&t=70s to hear a conversation with Whitley Streiber on the topic.

SUGGESTED READING

BOOKS AND WEBSITE BY DIANE TESSMAN – EarthChangesPredictions.com

GOD CLOUD – ALIEN AGENDA – EARTH CHANGES BIBLE– 7 RAYS OF THE HEALING MILLENNIUM – UFOS ARE THEY YOUR PASSPORT TO HEAVEN?

BOOKS BY WHITLEY STRIEBER

COMMUNION – THE KEY – MAJESTIC – THE SUPERNATURAL — BREAK-THROUGH, TRANSFORMATION

UnknownCountry.com – Website and weekly podcast

BOOKS BY BRENT RAYNES

JOHN KEEL: THE MAN, THE MYTH – EDGE OF REALITY – UFOS, TIME SLIPS, OTHER REALMS,

Free Monthly Newsletter - http://www.apmagazine.info/

UFOS DEJA VU

Since the passing of Anne Streiber, best selling author Whitley Streiber states that he still communicates with his late wife. After the publication of "Communion," the couple received letters from readers saying they had experienced close encounters in which their loved one had appeared on board a UFO.

Brent Raynes with Peruvian whistles which emit tones that might open portals.

Book details reports of contact with deceased loved ones aboard "space craft."

Cult producer Ed Wood seems to have gotten it at least partially right when he connected UFOs, aliens and the dead in "Plan Nine From Outer Space."

SECTION TWO
THE BRADSHAW RANCH

SEDONA'S HIGH STRANGENESS VORTEX

UFOS DEJA VU

UFOS DEJA VU

VENTURING ONTO SACRED GROUND – MY OWN PORTAL
By Timothy Green Beckley

PUBLISHER'S NOTE: I have long known that stargates exist and are not simply the work of creative science fiction writers, though I prefer to refer to these "exit ramps" to other dimensions as portals so as not to confuse reality with the novelization of the term. For 25 years I maintained a ten-bedroom "boarding house" with the most picturesque view ten miles outside of Glenwood Springs, Colorado, which I rented out to long term tourists and students from the nearby community college. "Beckley Manor" was located directly across from Mount Sophris, an ancient, once volcanic, mountain known for its twin peaks.

Said to be a "high energy vortex," New Agers had been known to plant crystals in the earth, and sacred drum ceremonies are frequently held there by Native Americans, as well as those wishing to pick up on the vibe. The area has also attracted an ongoing wave of UFO sightings. The objects – mostly in the orb category – have been observed many times by my associates Chris Franz and Jackie Blue, virtually from our veranda.

I should point out that the manor overlooked several hundred acres of pretty rough terrain, though there was a grassy knoll off to the side, which attracted grazing elk and bears that would rummage through our trash. Several psychics insisted that Sophris was a gateway to another dimension. To facilitate the sightings near my property, and thinking I might be present to see them sometime when visiting (which I never did), I went so far as to construct a rather crude stargate-like archway on the prairie a couple hundred yards outback, only to have the fierce winds blow it down repeatedly, until I finally gave up.

UFOs be damned!

It has often been said that Arizona is a virtual land of unabridged beauty and mystery which reaches out to those attuned to the vibrations of the landscape and the mystical draw of the state's many portals. UFOs are "old hat" here with the mesmerizing appearance of the Phoenix Lights seen by thousands, while the Superstition Mountains lure treasure hunters looking for that pot of gold beneath the ambrosial rainbow that never seems to end, stretching across the sky from horizon to horizon.

Let us not, in addition, neglect the fact that to the Native American tribes that inhabit the region much of the territory is on holy or sacred ground, pictographs on ancient canyon walls and inside caves proving that they have inhabited this area for centuries.

To prove our point, let us start out with the tale of an alleged vortex — i.e. stargate — revealed to us by our friends at mysteriousuniverse.org/ which is said to exist in the Southeastern mountains of the state which was discussed as only a minor part of their frequently issued blog, this one being titled:

Mysterious Portals and Stargates of the Ancient World

"In 1956, treasure hunters Rob and Chuck Quinn came to the area on a mission to find gold and lost Spanish treasures, but found something beyond their imagination instead. The two treasure hunters had already experienced quite a bit of bizarreness in the area in the form of mysterious floating lights at night, when they reportedly stumbled across a strange stone archway standing in the middle of nowhere, which measured 7 feet high and 5 feet wide and possessed columns of andesite 15 inches in diameter. The archway was oddly standing amongst scattered and broken geodes, their contents glittering in the sunlight. As curious as this all was, they had treasure to find, so they simply made a note of the location and continued on their way.

"When they later mentioned the archway to others, a local Indian guide known only as 'John' knew what it was, and had quite a few tales to tell about it. He claimed that on occasion people who had ventured through the archway had vanished into thin air, and that stones thrown through it would often not emerge from the other side. It was as if it were some kind of portal to—well—no one knows. John had also heard tales of the doorway glowing or shimmering, and he even had his own very weird experience there. He told them that he had once been trekking through the area and noticed that even though that day had been dark and cloudy the sky through the stone archway had appeared clear and blue, a phenomenon which he could not explain.

"The Quinns decided to go back to the enigmatic site, and things would only get weirder from there. At first there was not much to see. They experimented with the archway by throwing rocks through it and even shoving their arms in, but nothing strange occurred at all and their skepticism of John's fantastic tales grew. However, the following day as they rummaged about the site Roy and another team member witnessed the portal shimmering for several minutes like a heat mirage, and not long after this could feel a building pressure in their ears. This apparently lasted several minutes before dissipating. The effect could not be repeated, but a passing group of treasure hunters told them that the stone archway was indeed imbued with some sort of inexplicable force, claiming that as they

UFOS DEJA VU

had been camping there a rain of stones had pelted their camp from nowhere, and the stones had been warm to the touch. It was an intriguing story, but nothing else strange happened at the site during their expedition. They went away with more questions than answers.

"The Quinns would not forget the anomalous stone doorway, and in 1973 Chuck ventured out to the site once again. As he hiked up the canyon, he claims that when he stopped for a rest he noticed that there was another canyon that had not been there before. Thinking this to be rather odd, he climbed back down and entered the canyon from a different direction, where he realized that in fact he was in the same canyon he had been in before, only he had somehow been transported 250 yards down the canyon he had hiked along, and to another slope that was facing south rather than west. It was a very jolting experience that convinced him that he had traveled through some sort of portal and teleported from one place to another. It is all a very bizarre tale, and one wonders how much truth any of it has, or what became of this mysterious ancient gateway."

Treasure on Superstition Mountain is the second book in Elise Broach's Superstition Mountain series.
Have lost prospectors vanished into a vast portal there?

Publisher Tim Beckley constructed his own portal gateway on the prairie anticipating the arrival of UFOs. Unfortunately, they never arrived!

UFOS DEJA VU

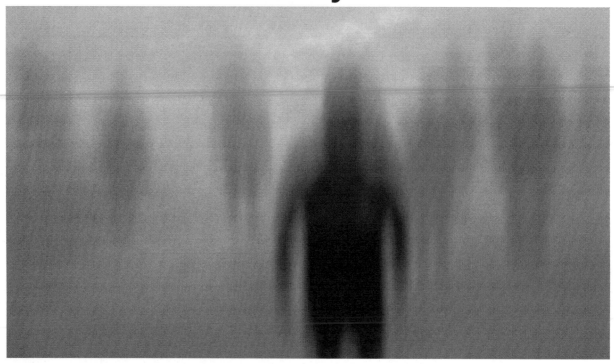

Thousands of people go missing each year as if they had vanished through a gateway or portal to another dimension

No one has yet come up with an explanation for the Phoenix Lights.

UFOS DEJA VU

THE FOUR VORTEXES OF SEDONA
By Timothy Green Beckley

Home to Tombstone and the OK Corral, Arizona is also the location of the Bradshaw Ranch which should be as famous, but has some catching up to do. This, despite the fact that the feds would like to think that, with their recent purchase of the land around the ranch, that the multiple phenomena taking place there would simply go away and never return.

If we thought about it, we could doubtlessly devote a thick book to the multitude of concealed mysteries associated with the Grand Canyon State. But we've more or less left that hefty writing chore to Tom Dongo of Sedona who knows more about unexplained phenomena transpiring here than anyone could possibly imagine.

TO MANY ITS ALL ABOUT THE VORTEX ENERGY

The proprietors of Sedona Red Rock Tours explain the beneficial virtues of the Vortex energy accordingly: "They – the vortexes – intensify everything they come in contact with and makes Sedona at large a huge amplifier. That means that everything you are feeling will be magnified. If you are happy while you are in Sedona, you may become euphoric or blissful. When you experience the sensation of love, you may be ecstatic.

"Conversely, if you are irritated, agitated or depressed, those feelings will also be magnified. It's for this reason it's very important to watch your feelings and mental chatter while you're in Sedona. You may experience instant demonstrations of just how powerful your thoughts and emotions can be.

"Vortex energy makes living here a constant learning and growth experience. We refer to this special energy field as the "Spirit of Sedona," a strong, creative, inspiring and loving feminine presence. And we consider ourselves to be here at her bidding, her will and her invitation."

A QUICK TRIP AROUND TOWN

At an elevation of 4,326 feet, Sedona is an Arizona desert town near Flagstaff that's surrounded by red-rock buttes, steep canyon walls and pine forests. It's noted for its mild climate and vibrant arts community. Uptown Sedona is dense with New Age shops, spas and art galleries. On the town's outskirts, numerous trail heads access Red Rock State Park, which offers bird-watching, hiking and picnicking spots.

And you know what? They're absolutely, one hundred percent right! If Manhattan is the Ying, Sedona has to be the Yang. In Chinese philosophy, Ying and Yang describe how seemingly opposite or contrary forces may actually be complementary, interconnected, and interdependent in the natural world, and how they may give rise to each other as they interrelate to one another.

Its estimated that Sedona attracts about four million tourists annually like a powerful earth magnet. The visitors come from all over the world to hike, rock climb, go to the local art galleries, and take back home some of the most incredible silver jewelry in the form of thick bracelets and squashes created by the various Native American craft people who live on or near the twenty reservations that dot the state.

One of the big attractions are Sedona's many vortexes—at least four, maybe more, but no need to count. Just go visit and experience them one by one. You can hike or take a jeep tour. I have been told that, "The subtle energy that exists at these locations interacts with who a person is inside. It resonates with and strengthens the inner being of each person that comes within about a quarter to a half mile of it." Another pinning proclaims that, "A vortex is a place in nature where the earth is exceptionally alive with energy. The term vortex in Sedona refers to a place where the earth energy swirls and draws to it's center everything that surrounds it like a tornado. At these magical sites, trees often exhibit this swirling or twisting of their trunks due the powerful vortex energy at the core of a Sedona Vortex."

A GUIDED TOUR

Red Rock Tours (928 282-0993) can take you to all of the popular vortexes, no problem. "The four best known Sedona spiritual vortexes are ," (they tell us) "Airport Mesa, Bell Rock, Cathedral Rock and Boynton Canyon. Each of these vortexes are well- documented and publicized. In 1980, a local Sedona medium named Page Bryant acknowledged and named some of the vortexes, causing them to gain in popularity as places for spiritual awakening.

"The vortexes are not new, however, and Native Americans including the Navajo, Yavapai and Hopi Indians long before recognized the energy and spiritual power of Sedona's vortexes, honoring the land in this area and using them only for sacred ceremonies.

"There are many Sedona spiritual vortexes than just these four. In fact, Sedona is enveloped with vortexes which have been detected and measured through dowsing."

A complete guide to each of the major vortexes can be found on line on various sites. We condensed what "A Guide to Sedona" had to say – aguidetosedona.com. This information is particularly useful if you are a trekker, especially if you are going on a long demanding journey by foot. That about sums up the ways and means we are talking about venturing around Sedona here. Me, I liked the Mustang convertible we once rented, but you can't make it up the rocky terrain without a four wheel drive. So looks are not everything in this case.

RED ROCK VORTEX – With so many scenic and energetic places in Sedona, Cathedral Rock Vortex stands out as one of the most scenic and most energetic places in Sedona! To feel the most power at Cathedral Rock, you must get to the Saddle of Cathedral Rock. But this is one of the most strenuous hikes and a dangerous/steep climb that is NOT FOR EVERYBODY! The hike/climb to the saddle is only 3/4 of a mile long but rises over 650 feet in elevation over the 3/4 of a mile. Only the fittest and most experienced hikers wearing the proper gear should even attempt to get to the saddle. For those of you who want to experience Cathedral Rock Vortex and not risk the Saddle Hike, then take the Templeton Trail to Oak Creek. Once you reach the creek, look for the Buddha Beach which is a flat area where visitors have built hundreds of Rock Cairns. We suggest you find a comfortable respite and relax to take in all the beauty of this magical place. Guides say this is also a vortex area and is certainly much safer than the hike to Cathedral Rock's Saddle area.

BOYNTON CANYON VORTEX – People really enjoy Boynton Canyon Vortex and it is one of the few vortexes you can reach out and touch. It's located just over one half of a mile from the Boynton Canyon parking area and very easy to enjoy. Many claim that there are actually two vortexes in Boynton Canyon. One is a tall red rock formation off to the east known as Kachina Woman and the other is an unnamed red rock formation off to the west. Many people believe these two actually form one vortex but because of the claims that people often feel energy at one and not the other it is up for debate. We believe that once you experience the energy at one of these spots, you are already on frequency with that energy and may not feel anything different when you visit the other spot.

BELL ROCK – You can't go wrong visiting Bell Rock Vortex because all of Bell Rock is considered to be very powerful and full of vortex energy. There are different ways to go about experiencing this vortex and there are many paths to take once you park. Parking is important here and we suggest you park in the North Bell Rock parking area (Directions and map below) because if you park in the south lot, you will need to hike 1 mile north because the South side of Bell Rock is too steep to hike or climb safely. Once you park in the North lot, hike past the sign board and head to Bell Rock Trail. Follow it about 1/10th of a mile or so where it will intersect with Courthouse Butte Loop Trail. You can either continue on Bell Rock Trail to explore the NE area of Bell Rock or turn right and follow Courthouse Butte Loop for 500 feet or so to a sign post on your left. Turn Left here and begin

UFOS DEJA VU

climbing up towards Bell Rock. Be sure to count the rock cairns and between the 10th and 11th cairn turn right and head towards the large flat rock shelf and you will see where you start to climb to Meditation Perch. It takes some doing to get there but the climb is worth the effort.

AIRPORT MESA VORTEX (most powerful?) – Airport Mesa Vortex is one of the most visited Sedona vortex sites because of its proximity to the center of town. Plus it is very accessible. To get to the top it does involve some vertical hiking and it is not for everyone, but the views from the top are stupendous in all directions. You will see fine examples of twisted trees (See pictures above) and many people claim to see colored orbs. This is truly one of Sedona's energy centers and will give you the vortex experience you are looking for. However, one of the prices you pay by visiting this vortex is that it is very popular and you are seldom able to be alone, especially on the weekends. If you get to Airport Mesa early in the morning at sunrise, the crowds are far less.

Sedona is known for its four main vortexes, though there could be quite a few more.

Eenie meenie miney moe - pick a vortex and lets go!

Jumping for joy, long time friend Sue Gordon gets the best possible view of Sedona.

28

UFOS DEJA VU

TOM DONGO AND THE TALL BLUE MAN
By Timothy Green Beckley and Tom Dongo

As we press on, it becomes obvious that Tom Dongo is "the big man around town." We thought we would start off our visit with a little anecdotal story of his.

"If you read the book, 'Merging Dimensions,' (now being revised) written by myself and Linda Bradshaw there is the story about a human like creature we saw on the Bradshaw Ranch in 1993. The creature was about 5'8," and every part of the humanoid being was a light brown color; Skin, shoes, clothes and all. We did not see its face as it ran diagonally away from us. There were three of us there at the time and we got a very good look at the being. It seems that it is the exact same type creature that was seen running in broad daylight in uptown Sedona. There are usually a lot of people in uptown, so no doubt a number of shoppers saw the creature. The man that filed the report said that in no way was this creature human. It ran through uptown and then across a back parking lot of a motel. He said no human could run as fast as this creature did. The creature stopped in the parking lot and for a few moments was eyeball to eyeball with the man that filed the report. He said its eyes were larger than a human's and were as black as coal. The creature then ran off and disappeared in seconds."

But nothing can be stranger than the ace sky watcher's tale of Sedona's "Blue Man." Kind of unnerving if you are not up to believing that the "visitors" are all around us, regardless if we can see them or not.

For a long time we have heard of "little green men" in relation to UFOs, if only in a knee slapping way. But I'll bet you didn't know that little—well I should say, "tall blue men" are in vogue, at least around Sedona, along with some other unidentified creatures observed from time to time. At least that's the way Tom sees it, according to what he told Charla Gene and I. The photo of the "Blue Man" was taken by Linda in his presence, and is intriguing for sure. "Linda's photo shows a left side head profile and the left shoulder of a man with blue hair," says Tom, adding, "I believe it is the first authentic picture of an alien of interdimensional origins."

Filling us in on all the lurid details, Tom added, "We didn't see the Blue Man when the photo was taken, but the figure is definitely there and he is very peculiar."

What is unusual about him?

"The blue haired man's ear is different from any human's. It is puffy and it has no ear hole. Instead, if has only two small creases along its outer edge. In the profile of the head the left eye is so small that it is not visible.

"Furthermore, there is an odd round mole like growth directly in front of his ear. Several MDs who looked at it said it almost appeared to be some sort of organ...maybe a secondary ear. Or it may just be a mole. One thing that really strikes me as uncanny is the angle forward V shaped side burn. Here is why. The side burn is almost identical to the side burns worn by James Kirk and the Enterprise crew from the original 'Star Trek.' There has never been anyone around here with that type of hair cut.

Tom takes it another step upon noticing that, "There are two fascinating aspects to this photo. One is that it was a bitter cold winter night in mid February, but the feet are in sandals and sticking out are toes that would be bigger and rounder than a humans. Another aspect is that the being or person was not visible to human eyes, but the legs cast an obvious shadow on the ground. I could not see the being but he, she or it had enough density to cast a shadow." How do you explain that?

We obviously can't!

Those who reside in and around Sedona who are of a metaphysical or New Age persuasion, seem willing to share with those who come here to meditate with crystals , and dream catchers and hunters of UFOs, how the town rests on several portals or gateways where all manner of odd events frequently occur. We're talking about entrances to an underground UFO base, a canyon where a flying saucer is said to have crashed, orange orbs that haunt a ranch once privately owned and which is now fenced off and owned by the government. Not to forget the above mentioned "blue man" who has appeared in photographs even though no one can see this presumed ultra-terrestrial, but his presence is hard to deny as he stands next to witnesses at a night time UFO skywatch and can be captured when you click the camera's shutter.

HE'S THE MAN ABOUT TOWN

Not wanting to be redundant, as we have reiterated there is no end to those who come from all over the world to attune themselves with nature. They take part in vision quests, meditate with crystals and dream-catchers and, above all else, look for UFOs.

And some of them are very lucky – some more than others. Perhaps one man above everyone else. .

Go to my YouTube Channel – Mr. UFOs Secret Files – and watch the video "In-

UFOS DEJA VU

visible Aliens 'Invade' Town – Underground Bases Exposed,"

https://www.youtube.com/watch?v=HFgvjjjrClM&t=1667s

You will get an agreeable, pleasant view of the town's supernatural underpinnings as experienced by a gentlemen who seems to be trailed by all that is unexplained and decisively strange.

Tom Dongo once ran the ultimate and most illuminating jeep tour of Sedona. He knows the continuum of this place like the back of his hand, as well as its incredible geology. And, if you wanted to go on a UFO Sky-Watch, you've picked the person who knows his alien lore—from the location of possible underground bases to secret canyons where UFOs have been known to hide.

On a visit to Sedona, when Tom came to where I was staying, he didn't arrive empty handed. Under his arm he had two hefty photo albums, the kind with glossy pockets that you can place 8.5x11 prints in for protection so that they don't get fingerprints all over them or have drinks spilled on the one-of- a-kind pictures.

Seated on the couch, my associate Charla Gene and I bombarded Tom with questions for an hour or more as he thumbed through the albums, selecting prime photos to show us undeniable proof that Sedona is "up there" when it comes to high-strangeness UFO cases. There were orange orbs, entities that showed up in pictures when they were not visible to the human eye and weird machines nestled back among the red rocks like something out of a science-fiction backyard. Sedona is known for its various vortexes, including one near the post office and one located on the road to the small community airport. Here again is an instance where the UFO shenanigans seem, at least partially, to do with a specific locale, i.e., Sedona, a known portal or UFO "window area," but is centered on one or possibly a number of UFO Repeaters as the primary impetus.

And while UFO sightings and encounters have not been confined to any one locale around Sedona (a tall, translucent being was even seen crossing the road in traffic a few years ago) one spot has generated more reports over the course of time than anywhere else in this community known for such natural landmarks as Bell Rock and Cathedral Rock. . . It is the Bradshaw Ranch!

We have mentioned the four plus portals in and around town but, boy, the Bradshaw Ranch, as we can see starting in the next chapter, has to be plum dab in the middle of some sort of "alien vortex" or all these paraphysical events could not, by any stretch of the imagination, be happening there.

UFOS DEJA VU

The scenes above are of the Bradshaw Ranch motion picture set with a sign claiming the property belongs to the United States. Visitors are forbidden after dark and guards have turned some people away. What's going on here?

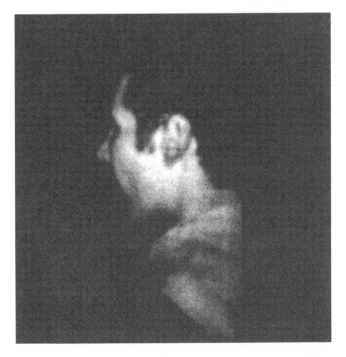

Invisible to the naked eye, the "Tall Blue Man" showed up in a photograph taken on the Bradshaw Ranch. He did seem to cast a shadow!

UFOS DEJA VU

Standing in front of a huge crystal on display in Sedona, Charla Gene bombarded Tom Dongo with questions about his many experiences in and around the town's famous vortexes.

Sedona's "portal jumping" skywatcher, Tom Dongo came to the author's motel room with two huge binders filled with the most outrageous UFO photos taken in and around town.

(Photo by Charla Gene)

Merging Dimensions by Tom Dongo and Linda Bradshaw tells the compete story of the Bradshaw Ranch and the undeniable fact that Sedona is a UFO gatway. The book will soon be available in an updated edition, if it isn't already.

"If you read the book, 'Merging Dimensions,' (now being revised) written by myself and Linda Bradshaw there is the story about a human like creature we saw on the Bradshaw Ranch in 1993. The creature was about 5'8," and every part of the humanoid being was a light brown color; Skin, shoes, clothes and all. We did not see its face as it ran diagonally away from us. There were three of us there at the time and we got a very good look at the being. It seems that it is the exact same type creature that was seen running in broad daylight in uptown Sedona.

"There are usually a lot of people in uptown, so no doubt a number of shoppers saw the creature. The man that filed the report said that in no way was this creature human. It ran through uptown and then across a back parking lot of a motel. He said no human could run as fast as this creature did. The creature stopped in the parking lot and for a few moments was eyeball to eyeball with the man that filed the report. He said its eyes were larger than a human's and were as black as coal. The creature then ran off and disappeared in seconds."

UFOS DEJA VU

THE BRADSHAW RANCH – A PORTAL OF PHENOMENAL DIMENSIONS
By Linda Bradshaw

PUBLISHER'S NOTE: Today the Bradshaw Ranch is locked up tighter than the proverbial drum. What used to be a grand tourist attraction in its "hayday" (hay for horses which are found on ranches, as opposed to heyday – well, never mind) has its gates chained and boarded up with a huge sign that reads – "U.S. Government Property – Trespassers Will Be Prosecuted To The Full Extent Of The Law!"

Though debatable, according to one well informed source, the Bradshaw Ranch "has seen more paranormal activity in recent years than perhaps any other location in the United States."

We're talking super portal to another dimension here. The mother of all gateways to God knows where. Sightings of creatures – visible and invisible – have been seen (or sensed) along with reports of all manner of craft, including floating and flashing lights, as well as a vast variety of colored orbs. Its like a Christmas tree of interplanetary ornaments.

History tells us it hasn't always been this way. Or maybe it has and everyone was just too busy to notice the perplexing shenanigans going on. The Bradshaw Ranch, all 140 acres of it, is located in Verde Valley about twelve miles outside of Sedona. It was purchased in 1945 by stuntman Bob Bradshaw who, for many years, ran the property as a film studio, one with a decisively western motif. I don't think "Gunsmoke" or "Wyatt Earp" was actually shot here, but you get the gist of what it means to be on the set of a pistol packing, six shooter, gun toting, "High Noon" imitation. Its a well known fact that Walt Disney's "Saucerers Apprentice" was shot nearby and someone back in mouseketeerland decided to rename it "Sorcerers Apprentice," no doubt because of Walt's close ties with the CIA and his being a supposed UFO disinformation agent.

But that's all conjecture, while what has been taking place at the Bradshaw Ranch all these years isn't by any stretch of the imagination.

It was around about 1992 when Linda Bradshaw began noticing that the family's westerly spread was surrounded by more than the beautiful red rock canyons and the star filled sky at night.

About one of her first encounters with a number of unexplained flying "balls of light," Linda says, "Before my eyes, a huge and brilliant light appeared in the sky above me. I did not see anything but the light itself, and it remained there for a few seconds." After several appearances, Linda began to carry a camera with her to record the strange activity which was increasing, much to her chagrin. It seemed that the balls of light were appearing pretty much in the same locale. She soon learned to act quickly when trying to photograph them. "I only had time to click the shutter twice, when the light instantly closed, leaving me to question whether it had been real." But the camera doesn't lie, as we like to say, and so she had the evidence she needed to prove to herself – and others – that there was something strange brewing on the ranch.

Little by little, Linda became convinced that the Bradshaw demesne was acting as a portal or a gateway to somewhere else. Other reports from "outside parties" also continued to mount up, such as this from the likes of Legendsatavist.com — Strange bright light blinking multiple colors west of Sedona near Secret Mountain. We witnessed a light that changed colors from blue, red, yellow, and white. We were on a dirt road near the "Bradshaw Ranch" and noticed as we were driving in its direction it slowly descended towards the horizon near Clarksdale. When we stopped to witness what type of movement it was making, it hovered for about 10 minutes. Then suddenly it zig-zaged west to east and back again. That's when I realized for sure that this was an unconventional object. It soon descended behind the mountain and was gone. It reappeared on the north side of town on our way back to Boynton canyon at much higher altitude and remained there for another half hour until we decided that it was time to go to bed and it was seen from our hotel door at the Days Inn in West Sedona. It was the brightest thing in the sky and had to have been witnessed by other people in town

But, let us now allow Linda, herself, to tell part of the story as taken from her own memoir and that of Tom Dongo's, "Merging Dimensions – The Opening Portals of Sedona."

x x x x x

LINDA'S "GALACTIC PARK"
By Linda Bradshaw

The daylight brings with it an incredible beauty. The majestic red rocks of Sedona greet the morning sun, and the rolling hills sparkle as the birds sing their praises to life. The beauty here can, at times, take one's breath away. I have the feeling that I have been transported to another place and yet, I am still here on Earth. How privileged I am to be here.

Then the nighttime comes and all is still. There is often the feeling of a "presence." This presence is not completely discernible, yet there is just enough to put a person on alert. The sounds come and our dogs frantically bark their protests to the intruders. At other times these same dogs cower in a corner and whimper.

UFOS DEJA VU

What happened? What or who is here? In time I would realize that some of these "presences" appear during the daytime also.

My first initiation to this strange world began one evening when I chose to step outside to witness a meteor shower. It was particularly visible and very beautiful, but several lights shot directly sideways rather than downward, and they had a different intensity to them. My curiosity was piqued.

The following evening I drove to the highest point on a hill entering our ranch, which gives a direct view of the canyon these lights appeared to originate from; coincidentally, this same canyon was rumored to hide a secret military base. This bit of information only heightened my curiosity further.

I parked my pickup truck so that I could look out the driver's window directly toward the canyon. With camera aboard, I arrived at 8:00 and approximately half an hour later I observed an airborne vehicle of some sort with flashing red lights. It was approaching the canyon from a westerly direction. There was no sound to this aircraft. Many planes had passed over by this time, at the same altitude, and each time I could hear their motors or jet engines. This emitted no sound. Just prior to reaching the mouth of the canyon the vehicle began a quick descent, straight down, and as it was about to go below the horizon, a white streak (almost like a meteor) descended perpendicular to it. The white streak was just to the right of the craft. I waited for the craft to rise again, but it did not. Then, approximately 20 minutes later, at a mountain peak farther to the right (and deeper into the canyon) I again saw flashing red lights with no sound. These lights hovered above the peak for about five minutes and then suddenly began an ascent almost straight up; it was not a gradual ascent as a plane would take. This craft was not a helicopter. All through this exercise I shot pictures of the strange flying object.

Next, I saw what appeared to be a meteor in the sky, but instead of being all white, it was a bright red ball shooting across the sky with a white trail behind it (similar to a comet tail). It was not at a high altitude, and amazingly it shot out of the canyon and into the sky toward the west. It then disappeared almost instantly.

I was in the process of absorbing what I had just seen when something caught my attention to the left of and a little behind my truck. There, hovering just above a bush approximately 20 feet away, was a large ball of white light (perhaps four feet in diameter) which would dim and grow small, then gradually brighten and grow larger. It repeated this again and again; a ball in the center with bright rays emitting out, then withdrawing and growing dim, all the time maintaining the small ball in the center. The pulse it emitted had the same cadence as a heartbeat.

The light was of an oyster shell color. I instantly perceived an intelligence about it. There was no question in my mind that this was a life form of some sort. It was hovering directly above a bush approximately four feet high.

This took me aback somewhat, and as I quickly discerned the situation, I wanted to leave, but I also wanted a photo of the light. So be it. I would snap one photo then rush out of there. The plan sounded reasonable to me. I raised my camera, focused, and suddenly received a telepathic message that shot directly through me. An unseen voice told me, "Do not take one more picture." After a chill ran through me from head to toe, I placed the camera on the seat, started my truck, and raced down the hill to our ranch.

What happened next was almost as bizarre. The following day I took my film into town to be developed. The shots I had taken were the last 16 frames of a 36-exposure roll of high-speed film. My husband, Bob, is a photographer and he had used the front portion of the film for an advertising job. Knowing I was planning to do some night photos, he had saved the last part of the roll for me to use.

When I later picked up the photos I noticed that only the interior shots that Bob had taken were in the packet. I went back into the photo lab and asked if they had possibly thrown out some film that they thought had nothing on it. The man said, "There was some unused film that we threw out. It was blank." I asked to see the film and he brought me a completely blank strip of film. There wasn't even an indication of frames, which always show up whether the picture comes out or not. When I asked about this he agreed that if there had been any shots taken at all, at least the frames would have shown up on film. I took the blank film home and discussed this with my husband, who then called the lab to ask if it could possibly be a length of film that just didn't get developed. He was told that was impossible, as the machine they use takes the whole roll of film and develops it automatically. The lab technician then again stated that the roll was simply not used, ending the conversation with the statement that the camera must not have been activated.

The camera I use is a Kodak 635 which is fully automated. So this is not a case of forgetting to open the lens cover; the camera will not work if the cover is closed. The camera had clicked and automatically advanced 14 times. I shot 14 photos. When I first arrived on the hill that night, the counter showed 20 shots had been taken. It was a 36- exposure roll. When I returned home, it showed that a total of 34 photos had been taken, leaving two unused frames.

What happened to the film? Perhaps it was due to some sort of human intervention. I even thought of magnetic fields around the area, which might have erased the film, but if that were the case, all of the photos would have been eliminated, not just the ones I shot. Sedona, because of its iron-rich red rocks, has strong magnetic fields.

Initially I was devastated, because this was the first activity I had documented in the area. Little did I know that in comparison to what was to come and what I would be allowed to film this setback was small change indeed.

UFOS DEJA VU

The Bradshaw Ranch began to receive national attention when Linda Bradshaw appeared on a popular cable paranormal program describing the ongoing activities taking place regularly on the property.

Since ranch was purchased by the government Keep Out and Warning signs have been posted.

The majority of events have taken place after twilight when mysterious orbs and creatures were known to appear.

UFOS DEJA VU

Readers looking for more Sedona UFO sighting reports can find 50+ sightings on www.Legendsatavist.com

The site seems to specialize in observation of orbs.

Have the feeling it may not be safe to walk down the streets of Sedona? Have no fear—this rescue vehicle could maybe save you.
Photo by Charla Gene.

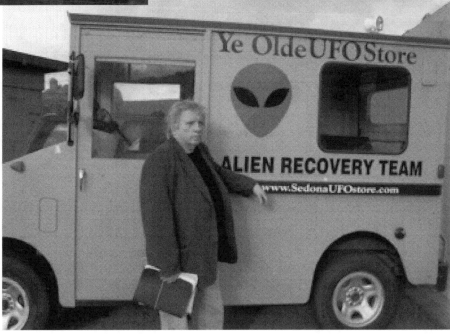

UFOS DEJA VU

THE 19 POINT SYSTEM
By Tom Dongo

PUBLISHER'S NOTE: When it comes to strange, unworldly phenomena there are certain aspects that repeat themselves over and over again. It's like getting trapped in a dream and you can't get out, and you fall asleep the next night and the same dream occurs. It doesn't matter what location you are in — what portal or vortex is in close proximity — you are bound to encounter a whirlwind of paranormal activity. You may have come to watch for ghosts, only to have to hide under the back seat when bigfoot comes a knocking. Or the dwelling you are residing in during the course of your investigations is haunted, and spooks are showing up everywhere. You see this sort of pattern over and over, as Paul Eno and his team mate Aleksandar Petakov demonstrate in our just published "Incredible Encounters," in a chapter about the Pennsylvania Triangle.

To start on our journey through the portals and vortexes we will attempt to explore (there are many more that will have to wait until a future date), we call upon our friend Tom Dongo once again. Tom, I should point out, is highly recognized as one of America's leading paranormal, UFO and ET researchers, as well as a remote viewer and wellness intuitive. For many years Tom has been a key lecturer at numerous conferences including the International UFO Congress.

He studied at the Berkeley International Psychic Institute in Santa Cruz, California and has spent decades in the greater Sedona, Arizona area, which is known as one of the premier UFO hot-spots in the world.

Mr. Dongo is the author of numerous books including, "The Mysteries of Sedona: The New Age Frontier;" "The Alien Tide (Mysteries of Sedona, Book 2);" "The Quest: Twenty-Six Who have Answered the Call to Sedona;" "Merging Dimensions Vol. 1, The Opening; " "Portals of Sedona, Everything You Wanted to Know about Sedona in a Nutshell."

We shall hear even more from Tom as we hike through the interdimensional terrain. But for now, we wish to concentrate on his findings as far as the repeating elements of our journey goes. Living in Sedona, Arizona, he has access to the four major vortexes around the town located in the red rock canyons of the state. In particular, we focus on his investigation into the events that have transpired on the Bradshaw Ranch at the heart of the pending mysteries we have been diagnosing.

TOM'S 19 POINT LIST

The following is a synopsis of what I personally have been involved in and have been researching here in Sedona, which I call my 19 point story. Almost all of this activity took place in one or more of the major portal areas.

For credibility purposes, I must mention that much of this activity, as bizarre as it might sound, has been witnessed by some of the most respected and credible citizens of the local area. In fact, several witnesses hold positions of such high responsibility, and authority that their credibility is almost beyond question. One out-of-town witness is in top management at one of the world's largest aerospace corporations.

To avoid embarassment to others, and as a personal polity, I never print real names unless specifically asked to do so and, therefore, I often get information that no one else gets because it's well known that I will not reveal sources unless asked or allowed by the sources themselves.

You will note that much of this information coincides with the information given by Linda Bradshaw, but is presented from the perspective of my own research notes and personal experience. It does not invalidate her perspective in any way.

The 19 points are as follows:

1. Frequent area sightings of moving balls of light, either silver, red, white, blue, or green. The sizes of these spheres have been approximately one to thirty feet in diameter. They have been seen to pulse at different intervals or glow steadily. The speeds of these objects varied roughly between 50 and 10,000 miles per hour.

2. Three reliable sightings of unknown animals, with no rational explanation of their origin. These animals were all four-legged, had either short or long fur, and their weight was between approximately 20 and 100 pounds. In mid-July of 1993 one such creature was seen by myself and a professional wilderness tour guide. We were in a jeep near Secret Canyon. This creature looked like a cross between a fox, a cat, and a raccoon. We got a very good look at it and we agreed that no animal like it existed on Earth. The animal was not a Coatimundi.

3. A number of strange tumors and inexplicable ailments found in rural domestic animals. And from two different horses in different places, a year apart, fetuses of four months each vanished overnight. Both horses were confirmed pregnant by a Phoenix veterinarian.

4. The appearance of highly unusual tracks. A number of six-inch wide, twenty-inch-long, three- and five-toed barefoot tracks have been photographed. I found in an out-of-the-way area one perfect track of a soft-soled boot. The print was ten inches long and five inches wide with an inward-bevel sole. There was no tread on the boot. On May 20, 1995, a longtime local tour guide, while on an outing with a group of hikers, discovered ten unusual footprints near Loy Butte. The footprints were approximately nine inches long, narrow at the heel and very wide at the ball. Each track had six toes. The tracks were in dry, dusty soil and were in excellent, well-defined condition.

5. Many reliable sightings of single and multiple military vehicles late at night and in the dark, early morning hours. One of these was a convoy of about 20 all-white, numbered tractor-trailers. In one single, isolated incidence, uniformed (probably Army) personnel were seen and photographed. Every single report of these military vehicles has them going into the canyons. There has never been one report of them coming out of the canyons. Where do they end up and where do they go into without being seen? We haven't the slightest clue.

6. About half a dozen cases of strong oscillating vibrations seeming to originate from underground. In one instance the witness, a psychologist, said the origin seemed to be several hundred feet above ground. When active, these vibrations are strong enough to shake large structures and be totally distracting to the residents involved. These vibrations are seemingly not of natural origin.

7. A number of cases of human abduction by aliens. One subject was hypnotized by an eminently qualified psychiatrist. The subject, a local man, remembers being taken into a spacecraft, seeing almond-eyed, large-headed alien faces, and having something done to his knees and temples, possibly implants.

8. Strange nocturnal movements of unseen, seemingly humanoid creatures or beings. The typical scenario involves hearing footsteps, with no creature or intelligent being visible. Doorknobs late at night have been rattled and turned by unseen entities or agencies. Screams and howls have several times accompanied these visitations. At least a half-dozen witnesses, including myself, have had living, invisible entities about eight feet in height brush past them. It is a sensation of warm, slightly electrified water brushing past. These incidents do not fit in with the usual ghost or poltergeist activity.

9. Domestic animals terrified by things unseen. In one case something in a large open area of about a half-acre could be heard hissing loudly at barking dogs that were penned up. The hissing creature was not a mountain lion. In one case a horse was found in its corral in a state of shock, and several thousand dollars were spent to bring the horse back to normal health. In a separate case, again in a remote rural area, a horse was found wedged into an angled tree. It was also in a corral and was apparently trying to flee something. Mountain lions, bears, and dogs can be ruled out in this incident and in most of the others.

10. A fire of extremely suspicious origin on top of Secret Mountain. Two thousand acres burned in August before the fire was finally doused. The fire was al-

UFOS DEJA VU

lowed to burn for five days before any attempt was made to put it out. UFOs and military helicopters and jet fighters were seen in the area at the time of the start of the fire. It seems to this researcher as if "someone" wanted "something" to burn and be destroyed - and perhaps it was.

11. Late-night (midnight to 4 A.M.) overflights as low as 100 feet of airliner-sized aircraft and especially military helicopters. One particular daylight case constituted outright harassment by two military helicopters. The two helicopters left when a twelve-gauge shotgun was pointed at one of the pilots by a very angry local man. One of the helicopters was hovering ten feet above this man's house, which is located in a sparsely populated rural area. Along with this there have been unrelated sightings of jet fighters and military helicopters looking for something on the ground. This has typically involved about a dozen crafts. In two totally separate, unrelated local cases military helicopters had, in the distance, "something" completely surrounded on the ground. Witnesses were too far away to see what the object on the ground was.

12. Three completely separate, months-apart reports of jet fighters and one helicopter flying at low altitudes and simply vanishing. In one case, a mirrorlike flash was seen on the midsection of a delta- winged jet fighter. The aircraft had clusters of rockets under each wing. There was a flash and then the aircraft vanished and did not reappear. Type of aircraft unknown - could be some sort of hologram experiment. One of these incidents, a vanishing helicopter, was eye-witnessed by a retired Air Force colonel who remarked, before being angrily dragged away by his wife, "I'll be damned. They finally perfected it!" Your guess is as good as mine as to what "it" is. The colonel and his wife left the area before they could be interviewed.

13. One midsummer 1994, 9 A.M., incidence of a reptilian-like humanoid running along a dry, rocky wash as though trying to elude someone or something. U.S. military personnel were reportedly seen in the area at the same time. The witnesses were two adults and one seven-year-old child.

14. Voices heard in the desert at night, talking in a strange, high- pitched, chattering language. I was witness to one of these occurrences.

15. Jet fighters flying with (escorting?) a sphere of white light as large as the wingspans of the fighters. This comprises usually two fighters and one ball of light. There have been at least a half-dozen of these sightings in the past two years in the Sedona area.

16. UFOs trying to look like aircraft or aircraft trying to look like UFOs. I have, on many occasions, watched large and small aircraft change running lights into different patterns in a seeming effort to look like a UFO or something else. Why? I have also seen what I think was a huge UFO trying to look like a large jet aircraft. The craft, however, made no sound whatsoever. A number of people have seen

44

that same occurrence in the Sedona area. I have also seen jet fighters flying in a defensive manner that, to me, strongly suggested they were expecting or were ready for an attack. An attack by whom or what in U.S. airspace? On the night of September 13, 1994, during a wild series of UFO sightings, I saw what looked to be a small, probably twin-engine plane coming toward me at a low altitude. With only the naked eye it simply looked like a small plane making an approach to the Sedona airport with its landing lights on. But when I picked up my binoculars and looked at it, it was a different sight altogether. In the middle, on the underside of the plane, was a red flashing strobe light. To the right and left of the strobe, and on what would be the middle of a light plane's wings, were two bright landing lights. But on the outer edge of each wing was a light that slowly pulsed with such a blazing brightness that at the peak of each pulse all I could see was the red strobe and a small inner portion of the "landing lights." It then turned east and flew right over Sedona. A UFO trying to look like a plane or a plane trying to look like a UFO? Whatever it was, it was not normal.

17. One instance of three military helicopters chasing at high speed, at rooftop level, two fast-moving three-foot-diameter balls of blue light. There were multiple witnesses to this event that transpired right across the city of Sedona at 10 P.M. on August 25, 1994.

18. Three incidents where jet fighters were chasing UFOs. In each case the UFO was a sphere of white light approximately 20 feet in diameter. In two days apart instances the UFOs turned and harassed the jet fighter by flying straight at it then making an abrupt 90-degree turn just before the UFO would have hit the fighter. Seems tempers might be getting a bit short.

19. Two more local incidents involving hikers or four-wheel drive buffs running into assault-weapon-toting soldier types and being turned back at gunpoint. This makes a total of about twelve of these incidents (that I know of) over a five-year period. These last two occurred on or about September 24, 1994. One was about a mile from the Sedona city limits and the other was in a remote part of the Coconino National Forest on the Colorado Plateau (Mogollon Rim), about 15 miles northwest of Sedona. In the typical scenario, hikers in a remote area run into two or more "soldiers" in charcoal-black uniforms bearing no insignias. The hikers, etc. are told in no uncertain terms that they are in a restricted area and are to turn around and leave the area the way they came. These soldier types in the black uniforms are almost always hostile, usually carry M-16 assault rifles, and give the clear impression that they will shoot if their suggestion is not carried out at once. They are not survivalist types or American Patriot Militia in training. Neither of these two groups has any idea of who the intruders in black are.

This 19-point saga continues in the area and in many other locations around the country. One really has to ask at this point - what is really going on?

UFOS DEJA VU

Tom Dongo and Tim Beckley on a Sedona vortex expedition along with Charla Gene (who snapped this photo).

< Tim Beckley and Raven De La Croix discuss Sedona portals on an episode of Exploring The Bizarre, archived here on YouTube with over 32K views.
www.youtube.com
watch?v=qRzFWWOm66Q&t=208s

Is this a paranormal timber wolf? Wolves have been extinct in Sedona for a long while, so what is this lovely animal doing there with glowing eyes?

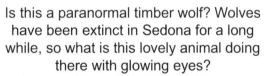

< Raven De La Croix, from actress to exotic dancer to leading edge "Goddess." Here she is with an orb hovering above her head, a common occurrence at her Sedona vortex headquarters.

http://ravenscosmicportal.com/UFO.html

UFOS DEJA VU

THE RANCH ON THE EDGE OF FOREVER
By William H. Hamilton III

PUBLISHER'S NOTE: I have known Bill Hamilton it seems like forever. He has a military security background and has held an interest in UFOs since the early days of contacteeism in California, when thousands would venture out to the sand dunes waiting for the saucers to land around the very famous New Age landmark, Giant Rock Airport in the Mojave Desert. Bill is a qualified researcher (acting as the Executive Director of Skywatch International, Inc.) as well as an experiencer who believes he might have been abducted near Area 51. He has written several books for our Inner Light – Global Communications such as "Cosmic Top Secret," "Alien Magic" and "Time Travel Now." When it was still being printed, he also contributed to the nationally distributed newsstand publication "UFO Universe" which I packaged for eleven years, before the bottom fell out of the magazine business. A little while later, Hamilton III retired from his active involvement in UFO research and moved to middle America with his wife Pamela to engage full time in more lucrative business activities.

I believe in my heart, that Hamilton first clued me in on the activities on and around the Bradshaw Ranch. In fact, he sent me an article, which I was delighted to publish, describing his experiences on the property, from which we can take an excerpt at this point to show how hard core the phenomena was on the Bradshaw Ranch at one point in the 1980s: We commence with Bill Hamilton's engaging report:

* * * * *

There are places of mystery in the world. Strange things are seen and bizarre events unfold in these mysterious places. Sedona, Arizona is such a mysterious place. Visitors flock to this Arizona New Age capital to view the striking red cliffs and rocks that highlight the verdant growth of pine and juniper as well as to feel the uplifting energy of Sedona's vortices. Conservative people think the vortex believers are a little short of a full deck. After all, if the vortex is invisible, how does one know that a vortex is on one site and not another? Even if people claimed they could feel the energy of the vortex, why doesn't it just twirl them around?

47

UFOS DEJA VU

There are other visible mysteries that occur in the Sedona region. There are UFOs and strange orbs of light that move through the back canyons, notably Long and Boynton Canyons. A mysterious military or paramilitary presence has been reported in Boynton Canyon, Secret Canyon, and Sycamore Canyon. Back in 1991, I traveled to Sedona to visit a man who lived in Long Canyon. John was a caretaker on an old housing project. He had sighted a large boomerang traveling from the direction of Secret Canyon in the north to the edge of Sedona in the south. The remarkable thing about this very large vehicle was its method of flight: it was flying on end and leaving a sparkling trail!

Strange, unmarked helicopters had been seen coming from the direction of Secret Canyon and flying south, the same direction as the boomerang flew. That day when I was in one section of Long Canyon I could hear the whopping blades of a helicopter behind the mountains to the north. Shortly, an olive drab helicopter flew low over the trees after taking off from somewhere within the canyon. Once, a fellow investigator tried to hike back to Secret Canyon around one o'clock in the morning when he heard the whopping swish of helicopter blades. Attempting to follow the trail back into the canyon, he was stopped by a voice that emanated from a loudspeaker warning him to go no further. Thinking the speaker was there to ward off hunters, he proceeded further when he was stopped again by a laser-targeting light moving around his chest. From the direction of the laser's source, he heard another voice telling him that he had entered a restricted area and that he was to turn around at once.

Field investigator and researcher Tom Dongo has spent his last few years living in the Sedona area, interviewing witnesses, and chasing UFO sightings, abductions, and other paranormal phenomena as well as trying to track down the reports of a secret military presence in the canyons. He has written books on his findings. He kept searching for a focus for the phenomena and found three; one was on a 90 acre ranch off the Boynton Canyon back roads in an isolated canyon between Red Canyon and Loy Butte.

Our friend Kim Carlsberg (renown photographer and author of a vital book on abductions, **"Beyond My Wildest Dreams, Diary Of A UFO Abductee"**) had moved to Sedona to raise her children and attend Northern Arizona University in Flagstaff. We introduced her to Tom Dongo and she started going on nightly sorties with Tom out to the ranch in this isolated canyon where all manner of things were happening. Tom always brought his camera loaded with 400 ASA film and a time exposure trigger. Kim brought her own camera. They would shoot one, two, or three rolls of film just pointing and shooting whenever they felt moved. The owners of the ranch are Bob and Linda Bradshaw. Bob was a freelance photographer as well as a rancher and his beautiful photos of Arizona landscapes have been published in books and magazines. Linda seems to be the contact point between worlds. Her experiences of the paranormal go back to childhood.

UFOS DEJA VU

My wife and I were invited to go out on a sortie and meet Linda. We gathered our cameras, videocam, binoculars, and tri-field detector and headed north for Sedona on October 7, 1995. I was prepared to meet the unknown.

We rendezvoused at Kim's house in Sedona. Tom arrived, bringing photographs that he had taken at the ranch. These photos mostly showed a variety of inexplicable light phenomena. In one photo, a randomly laced trail of varicolored light hovered over the ground. Other photos showed light streaks, orbs, and, in a few cases, structured objects. Tom did not see most of these lights and orbs, but they registered clearly on his film. I was eager to travel out to the ranch site before sunset so I could get a look at the surrounding terrain.

We traveled in two vehicles, Tom in his van, and the rest of us in Kim's transport van. We piled jackets and cameras in the back. The road was rough and rocky in places and riding in the van gave me the feeling of riding in a nineteenth century stagecoach. The last mile or two along the ranch road was the roughest surface we had yet encountered. We managed to arrive as the sun was setting and I could still get a good daylight view of the property. The main house and two other buildings sat squarely at the bottom of a little valley surrounded by trees and bushes. We were far away from city noise. Linda and her dogs came out to greet us as we disembarked from the vans, making us feel welcome and invited. We went into a well- decorated house that had all of the accouterments one would expect to find in an urban house and sat around the kitchen table. Linda served us coffee and handed us a large photo album filled with the strange pictures of paranormal light phenomena. The sheer number of such photos was remarkable. Others have taken similar photos, but not in such quantity. Linda told us parts of her story as we sat looking through her fascinating photo album.

Linda tells us that she has experienced strange phenomena all her life, but after moving onto the ranch, the frequency and strangeness of the experiences increased. One night while she and her husband were sitting in the kitchen they heard a noise that sounded like shattering glass come from inside their kitchen space, but nothing was seen and no shards of glass were ever found. She has heard footsteps both outside and inside the house when nobody could be seen. To further compound the mysterious events, some unknown agency would lift her camera at night and snap pictures around various parts of the house. One picture showed the light over the kitchen table, yet the area surrounding the light was completely dark! Another picture taken by the unknown agent showed the yard as seen through one of the kitchen windows, but it wasn't quite the same yard as a portion of the scene seemed blocked off by a mysterious illuminated border.

As it was getting darker and a full moon was rising in the east, we decided to gather our coats and equipment and go for a walk around the property. Linda first took us to the horse corral. There were many nights when her horses were spooked by something that left tracks in the dirt. These large three-toed tracks were not identifiable as belonging to any of the known wildlife in the canyons. Linda indicated a trail that we could walk toward an old western movie set that had been erected sometime ago. She said that we would find more tracks as we walked into the old west town. Most of the buildings were just facades. Linda said we should look for the tracks of Big Girl, a bigfoot that both she and her son had seen on occasion and which had left evidence of its presence at nearby locations.

I was told that if I tripped by camera flash in the dark I would see that the air was filled with an unusual sparkling energy. I deliberately flashed my camera in the dark while looking through the viewfinder, and, indeed, I could see glowing particles filling the air. Seeking a conventional explanation for this spectacle, I surmised that particles of dust must be suspended in the air, and that my flash simply reflected off these numerous particles of dust, but as I continued the experiment, I noticed that the glittering particles did not move around as I have seen dust move, but rather seemed to be suspended or frozen in the air.

As we continued walking, I noticed from time to time, a tiny light would flicker on in distant bushes then extinguish. I asked Tom about this, and he said that he noticed this on all of his frequent excursions out there. They appeared to be like the fireflies I used to watch at my uncle's house on Long Island when I was a young boy. But fireflies were not known in Arizona.

Tom kept looking around the horizon for the lights of UFOs. He claims that the frequency of UFO sightings over the canyons had picked up lately. He then proceeded to pull night-vision binoculars out of his pack to scan the horizon more closely for signs of movement. I could see a number of aircraft lights off in the distance to the south and east of us. When we came to the movie set we turned our flashlights on the ground and could make out some unusual tracks. One track looked exactly like those I had seen published of alleged bigfoot tracks. I photographed it. Other tracks were of the large three-toed variety akin to those made by large birds or small dinosaurs.

At the end of the trail sat a platform made into a gallows. It was an ominous sight to be greeted in the light of the full moon. The women climbed the stairs to sit on the platform, followed by one of Linda's dogs. Tom and I walked around, scanning the horizon for any faint movement. The air was still and silent. Occasionally, Tom would raise his camera and shoot on a seeming whim. When the camera flashed, I could see the strange glitter in the air, but nothing else seemed out of place or mysterious.

After standing around for a half-hour or so, we headed back to the ranch. When we all got cozy in the kitchen, Linda served coffee and I started to ask more questions. Linda told me about her son Victor's experiences. Victor had taken a hike back in the canyons one day and stumbled upon some white trucks and men in white suits. One flat-bed truck carried a wingless craft on its bed. Victor felt that he was being observed. He had his video cam at the ready and scanned the scene below his position when he thought he heard a noise in the bushes behind him. When he turned to look, he saw a creature peering back at him, then duck out of sight. This creature looked like a typical gray alien. On another hiking expedition into the canyons, he had become lost. He was walking in one place in daylight, then he found himself in an unknown location and it was suddenly night!

Tom had also seen a creature, but it was right near the ranch. One night while Kim and Linda were talking, he saw a small humanoid in a brown suit dash across the field near the juniper tree. The dogs gave chase, barking and running after the entity. As it neared the fence line, it seemed to vanish into thin air, and the dogs quite suddenly gave up the chase. Tom had never seen anything like this and was startled only after the entity vanished from sight.

Linda gave me a copy of the book that she and Tom had published, titled "Merging Dimensions." It is replete with dozens of photos. Unfortunately, the photos are reproduced in black and white and one misses seeing some of the startling colors that are visible in the original photos. Each of them are accumulating more anomalous photos every week. Anomalies have also appeared on some of Kim's photos.

Pamela and I went back to the ranch for a second visit on December 7, 1995. The night was even colder than the first night's visit. After nightfall, the stars sparkled like jewels. Tom set his camera up on a tripod with a timer attached to the camera to get long exposure photos of distant moving lights. Tom and Linda had been seeing "a ship" that would characteristically pop up behind some far hills to the south and west. Pamela had loaded her camera with 400 ASA film. We did see some unusual moving lights that night, but they seemed too far off to classify them as UFOs. Sky watching is a game that involves a lot of patience.

During a break in our watch, I went inside to get warm and convinced Linda to show me the video that Victor had pieced together from several different shoots around the canyon. This fifteen to twenty-minute video was shot in daylight and darkness. On it I saw some of the most peculiar "things" flitting around near the ground. One segment focused on a low-flying airplane which was blocked out by a brilliant orb of light that flew across the plane's path. The video was compelling. Very strange things were happening on this ranch.

I kept asking Linda details about the photograph of the window or gateway taken near the juniper tree. She had only seen a bright rectangle of light, yet the developed photo looks like a window into another world. Unless Tom, Linda, and now Kim were in collusion on a magnificent hoax, I would have to think that the photos are untampered and show what they purport to show. While it is true that they have not been subject to expert analysis, it still leads one to speculate about the happenings on the Bradshaw Ranch.

UFOS DEJA VU

There seem to be three significant elements to the Bradshaw events: 1) The Gateway Window; 2) the ever-present orbs; and 3) the sparkling energy in the air. The presence of humanoid entities, mysterious craft, and mysterious animals seemed tied into the elements offering clues to the mysterious world that was intersecting our familiar world in this remote canyon beyond Sedona.

Looking at the Gateway photo reveals a number of things. The scene inside the window area is in daylight. A structure that looks exactly like a telephone pole with the cross struts and insulators is on the right. A third of the way down from the top of the window one can see a dark oval object that looks like a flying saucer. At the bottom left corner is a jumble of indistinct objects that could be foliage or even a humanoid figure. What is even stranger is an embossed area to the left of the phone pole that looks like a light round object with projecting lines and the number 39 raised up to the right of the round object. There are two photos of the window taken in rapid succession before the window closed down. There are no telephone poles on the ranch property and there are certainly none stuck in the ground next to the juniper tree. If one wanted to come up with a hoax scenario, it would have to involve snapping a picture in daylight and superimposing it over a picture taken at night in another location. Of course, then one would have to explain how the daylight photo was taken of a hubcap or some other plain object thrown into the air to simulate a flying disk. Photo experts could examine the negatives for signs of double exposure or superposition. I, for one, would like to see the photos analyzed, but for the sake of argument, let us examine the implications if the photos were taken as described by the witness. Additionally, supporting the authenticity of the photos is the fact that photos taken by Kim and Pamela also show anomalous images.

A little research on photographic film reveals that it is essentially a thin plastic base coated with an emulsion. The emulsion is composed of gelatin within which tiny particles of light- sensitive salts have been suspended. Because of the silver halides in the emulsion, all light-sensitive photographic emulsions are sensitive to the blue, violet, and ultraviolet end of the visible and invisible spectrum. There are emulsions that make film sensitive to infrared frequencies and, of course, film made sensitive for x-rays. Objects beyond the range of sight could register on various types of film. No special film or filters were used by Tom or Linda to capture the invisible energies around them.

Some percentage of abductees have reported seeing glowing orbs. These orbs will appear in the abductee's house and pass right through solid barriers. In Brian Scott's case, he reported seeing an orb pass in front of a dresser mirror without reflecting in the mirror. Others have reported seeing orbs float above the ground without reflecting light off the ground even when the orb appeared brilliant to the observer. Orbs have appeared of various colors and from baseball to basketball sized. An orb is a good candidate for a hyperdimensional object. Not all of the orbs photographed are visible, yet they appear on film as translucent objects.

Even when orbs are visible, they are capable of passing through a solid wall or closed window. Their light emission may excite the optic nerves in humans and animals, yet may not reflect off surrounding surfaces. In actuality, the light-band frequencies of radiation may pass through a material object because they cannot be fully absorbed or reradiated by atomic electrons, and yet their radiation will affect film because of radiation at ultraviolet frequencies. The impression made on the film shows variations in photon density between different orbs appearing in the same image. We can only surmise that there exists stages of progressive materialization of hyperdimensional objects where they can appear only as a ghostly glow or become a fully objectified source of light. The evidence viewed in this way gives strength to the idea that the hyperdimensional parameter is a frequency dimension and not an extra-spatial dimension.

In other words, the orb is moving into and out of this dimension by shifting its frequency and perhaps its phase. According to physicists, subatomic particles are composed of wave packets, and these matter waves can pass through a small slit and show evidence of wavelike interference. Most of the matter waves we can see absorb and reflect photons within a narrow band of frequencies. Suppose objects composed of higher frequency wave-packets, and out-of-phase with our normal mundane sensible world were to approach a mundane material barrier composed of everyday protons and electrons. It would probably look just like a ghost passing through a wall. The wall would offer it little or no resistance. If a hyperdimensional cesium atom would be compared to a mundane cesium atom in normal dimensions, we would probably measure a faster rate of time and the space the cesium atom passed through would be isolated and separate from our normal space. There could be various levels of dimensional states. Without a more rigorous data collection and analysis, this remains only a speculative hypothesis.

We might also consider that the "glitter" seen in the air around the ranch property is caused by radiation from another dimension causing excitations in atmospheric molecules. Some of the molecules of oxygen and nitrogen could be absorbing radiation from the other side of the dimensional barrier, causing random photon reradiation in the air. The "glitter" appeared static. These glittering particles did not seem to be moving.

The implications of this phenomena require a reassessment of the nature of physical reality. If our scientists are studying only a portion of physical reality and basing theoretical models on the data collected from only one frequency dimension of the universe, then we may be harboring a very narrow view of this all-encompassing universe. Other worlds peopled by other beings might not only be found in the far reaches of intergalactic space, but in the near reaches of extradimensional space.

A phenomena that resembles the orbs is ball lightning. Ball lightening usually appears during thunderstorms. The ball is usually bright, fuzzy, and orange-yellow in color. Ball lightning has been seen that was up to 16 feet in diameter. Ball lightning has been seen to materialize inside enclosures, even an all metal aircraft. How it accomplishes this feat is unknown. If we could produce ball lightning in a laboratory, we could probably experiment and record its weird behavior and antics and gain some insight as to how the orbs materialize inside of homes.

The typical view is that a parallel reality exists along a fourth-dimensional spatial axis, however it is possible that a parallel reality exists along a fourth-dimensional temporal axis or energy axis. If space and time are fundamental dimensions, then it is possible that universes exist in other dimensions of space and time. The real hint comes from the energy expended in moving from a point in our world to a point in some adjacent world. But until some scientists take this phenomena under serious study, we are not likely to demystify these bizarre happenings.

Some UFOs may be someone else's spacecraft from an extra-solar planet in our Milky Way galaxy, but some UFOs may be visiting us from points of space that cannot be seen. I remember a conversation that I had with my informant, Charlie, about objects that had completely disappeared from the surface of the earth, some which he acknowledged were known about, and others that were classified as military secrets. According to Charlie not all UFOs come to earth from outer space. Some tunnel here from another space-time dimension. He told me that some of the recovered crashed disks had a unit that we called a transpatial resonator. These resonators could open portals allowing the craft to pass from one dimension to another. Maybe Charlie is telling the truth. After visiting the ranch on the edge of forever, I not only think of Sagan's billions and billions of stars, but the possible billions and billions of universes that are out there somewhere.

x x x x x

I think we should point out – or reconfirm – that Linda sold the ranch some time ago and when last heard from now lives in Montana. Breaking News! – The Bradshaw Ranch is now leased from the forest service by the University of Northern Arizona for native plant research. They will build a dormitory where the ranch houses were.

That does not mean, however, that Tom Dongo is hanging out doing nothing. Though retired from the tour business, he hasn't lost interest in the topic of UFOs, portals and the paranormal in general. Nor has he forgotten about his many experiences about the Bradshaw Ranch which is detailed in **"Merging Dimensions,"** an updated version of which is due out soon (check Amazon).

UFOS DEJA VU

William Hamilton III and wife Pamela seated at one of our UFO conferences.

Strange rock with unknown inscription found in dry creek in Sedona Canyon.

Bell Rock is known as "UFO Corridor." >

< Group shot taken at UFO skywatch in Sedona. Note eyes of "alien creature" behind individuals on right side. The being was not visible at the time the photo was taken.

UFOS DEJA VU

Tom and Charla Gene while on UFO expedition in Sedona with Tim Beckley. For more UFO photos and true tales of Sedona visit Tom Dongo's site: http://tomdongoufoparanormalblog.blogspot.com/

Bill Hamilton is one of the most famous and longest serving American UFO researchers.

Before he retired, he worked as a Senior Programmer-Analyst at UCLA. He has degrees in psychology, physics and IT.

From 1961 till 65, he worked in the United States Air Force Security Service.

Bill had his first UFO experience back in the 1950s.

Author of 8 books on UFOs

UFOS DEJA VU

VORTEXES AT PLAY – SYNCHRONICITIES AND A POSSIBLE HEALING
By Timothy Green Beckley

I am often asked which portal or vortex do I think is the most powerful, the one that attracts the most interdimensional phenomena? The answer to that question would have to be conjecture. Regardless if its the Bradshaw Ranch, the Skinwalker Ranch, or the phenomenal events taking place in the San Luis Valley, (or one of the other places mentioned in this volume), the phenomena may ebb and flow. Sometimes all can be quiet for a long period, even years, only to flare up at some future point in time. It is generational in that, once reported it never completely goes away, but the phenomena may be in hiding—say for example, if the property is taken over by someone with a mental "do not disturb" sign out. Or the gateway is not open on the other side for the parties over there to gain admission to our dimension.

Personally, I know that the portals in Sedona are powerful ones. I have two very heart stopping synchronicities in an area that is in close proximity to the Airport Mesa Vortex. The synchronicities – dramatic ones – took place at the exact same location which happens to be the Coffee Pot, a well trafficked restaurant just as you come into town.

But let me explain more fully. The majority of researchers, I would have to say, see these widely scattered portals as having a negative effect on our history, culture and society. But negative to whom? By what standards? We cannot justify the morality and behavior of the Visitors by comparing it to our fundamental beliefs. I suppose you could say they are guilty until proven innocent.

For a long time Tucson has been pretty much my second home. On a regular, usually an annual basis, I get together with my UFO chums Charla Gene, Christine Franz Dickey, Ed Biebel and Allan Benz, who is a real "ole timer" on the scene, having been involved in UFOlogy for a respectable number of decades. Currently, Allan is the director of the World UFO Group and the U.S. organizer of World UFO Day, which celebrates the birth of the modern UFO era which began with Kenneth Arnold's sighting of nine crescent shaped "wings" over Mount Rainier on June 24, 1947.

UFOS DEJA VU

A couple of times I have participated in meetings held by Allan for Tucson's "inner circle" of UFO devotees. What has always intrigued me about Allan is his background as librarian for the now long-defunct international organization – the Aerial Phenomena Research Organization. Best known simply as APRO, it was at its zenith as the second most popular UFO group in the world, next to the National Investigations Committee on Aerial Phenomena (NICAP). Unlike the Capitol Hill-based NICAP, APRO was caught up less in "officialdom," had no political affiliations, and was more interested in who flew the saucers than whether the government was trying to hide the truth about UFOs from the American public.

They had no CIA operatives on their board of directors. Instead they pulled in correspondence from a global team of researchers who studied landing cases and reports of close encounters with humanoid beings gathered from Italy, France, Sweden, and even Russia, which in those days was a real effort, since the "Reds" had a very closed society, especially when it came to discussing UFOs and space beings. The state run Soviet press ridiculed the subject, if they mentioned it at all, going out of their way to identify the unidentified as purely a capitalist phenomenon, contrived as a way of taking people's minds off of troubling social issues taking place in American society. APRO was run efficiently by the husband and wife team of Jim and Coral Lorenzen, who also edited a nicely done, very professional (for that period at least) newsletter, the APRO Bulletin. Allan Benz was voted in as official librarian (he held a degree in that profession) and, as such, his job was to catalog and file the hundreds of translations of reports streaming in from all over the world so that they could be pulled up quickly in an era when Kindle tablets, iPhones and lap tops still were quite a ways off.

Believe you me, I sat spellbound in his apartment as he told me how Steven Spielberg's production company had gotten in touch with APRO, requesting that they act as consultants on a forthcoming project, which turned out to be the making of the fabulously successful "Close Encounters of the Third Kind." Jim and Coral turned the consulting job over to Benz who spoke to representatives of the Hollywood-based production company numerous times, providing them with the material they needed. Speilberg, if you recall, had his own UFO encounter as a teen and was very much into the subject. Ironically, I became an adviser and editor for the "Close Encounters Poster Magazine," a glossy pullout publication with hundreds of thousands of readers. My job was to compare the dramatized UFO events in the movie to actual UFO sightings taking place all over the world.

After spending a good five or six hours in the company of APRO's ex-librarian we said good-bye to Allan Benz and headed down the road toward Sedona with Charla Gene who is a comrade in UFOlogical arms and my official photographer when I am in the area. We told no one we were headed out of town down the highway to do a bit of ghost hunting in the haunted town of Jerome and then venture further on to mystical Sedona.

UFOS DEJA VU

HEADS UP WHEN ENTERING THE SEDONA MATRIX

The drive from Tucson to Sedona is long and hot, especially during the summer. We're talking six or seven hours with the air conditioning blasting. About ten miles outside Tucson the traffic thins and in no time you are out in what we back east would call the "boonies." When the terrain allows, and you're not lumbering along some twisting mountain route, the road whizzes by as you put the pedal to the metal, which in my case is provided by Charla, behind the wheel.

On the way to Sedona we chatted about entrances to an underground UFO base, a canyon where a flying saucer is said to have crashed, strange orange orbs that haunt the terrain, and for what reason the area is home to at least four vortexes.

COFFEE AND OMELET NUMBER 103 PLEASE

Now, I know some people claim they can get pretty ripped on an ordinary cup of java. Take it black without any sugar, and God knows what the caffeine can do to you. Frankly, I can drink coffee before going to bed, and it doesn't make me toss or turn any more than I would normally. On your way in or out of Sedona, I would consider it "bad manners" not to stop between the hours of 6 AM and 2 PM at the Coffee Pot Restaurant and order yourself some breakfast, even if you normally would be having lunch. The Coffee Pot is known as the House of 101 types of Omelets, and believe me when I tell you I'm not kidding that there is something omelet-wise on their extensive menu to satisfy any craving. Order up Omelet #34 if you like Avocado, Onion, Mushrooms and Turkey Gravy (yum?), or better yet Omelet #101 (the last on the menu) consisting of Jelly, Peanut Butter and Banana – that's my favorite, oh boy!

WHAT IS THE SIGNIFICANCE OF 2012?

Furthermore, I know you're going to laugh at me – and a goofy chuckle is justified in this case – when I tell you there is something damn mystical, something very strange, about the Coffee Pot. Could be its location, being that it is located on AZ Highway 89 at the entrance to the Coconino National Forest and just a scant quarter of a mile or so from the New Age focal points of Cathedral Rock and Bell Rock. It is positively a beautiful setting at sunrise or sundown in anyone's estimation, and a high energy focal point if you want to charge your chakras. And, by the way, it is situated almost on top of the Airport Mesa Vortex.

If there is a New Age trend in the wind, Sedona's metaphysical party line operates overtime. If it's teachable, you can find a place in town to take classes in it. If it's a new product, they will likely have it first at the Center For The New Age or Crystal Enlightenment.

I'm not sure when all the fuss about 2012 started. But everyone – almost everyone anyway – said they felt a change was a brewing, that a big event was going to shake up humankind on a specific date – that date being the 21st of December

UFOS DEJA VU

2012. This day was said to be the end date of a 5,126 yearlong cycle in the Mesoamerican Long Count Calendar as specified by a group of Mayan priests, part of a hierarchy of Shamans whose culture extended across Mexico, Guatemala, Honduras and El Salvador. That hierarchy still exists to this day in varying degrees.

It was believed among those who prophecies that various astronomical alignments and numerological formulas would be in conjunction on that day that would mark the start of a period during which Earth and its inhabitants would undergo a physical or spiritual transformation Thus 2012 would mark the beginning of a new golden era of peace and harmony on the planet.

Some negative soothsayers said the world was going to come to an end and that the mysterious Planet X was going to pop out from behind the Sun and clobber Terra Firma, spinning us out of orbit and crushing us to death while depleting the atmosphere of Earth. It was, to me anyway, a sort of updated version of the Harmonic Convergence that had been celebrated back in 1987. If one had to pick a song that would best exemplify what we are talking about, it would be "Age of Aquarius" by the Fifth Dimension.

For a bit, it looked like things might get a bit dicey in Sedona on or around the day in question. There were plans for thousands to descend upon the town for a mass meditation and prayer to the heavens. A linking of hands across the galaxy, so to speak, was in the works. Who can have anything against such a concept? Well, there is a segment of those who live in Sedona year round who hate this sort of attention. They don't see anything particularly mystical about where they reside, and to their way of thinking the "crazies" should all keep their distance.

As we know, 2012 came and went and Earth didn't change vibrations as far as we could notice, nor did anyone I know ascend into the heavens. For all intents and purposes, December 21 passed us by without a hitch. We were not sucked through a wormhole, nor did Planet X arrive, transporting with it the giants known as the Anunnaki from the planet Niburu, said to be coming here to steal the planet's gold resources and make sex slaves of our women. I could have told everybody that nothing spectacular was going to happen well in advance – these sorts of predictions seldom come to fruition.

But, then again, I did have a personal synchronized experience around that time – actually two of them – that I have been trying to build up to for the last few thousand words or so. Both experiences took place at the Coffee Pot, if that isn't weird by itself. I was having brunch with Charla and offered to pick up the tab since she had done all the driving and had even pumped the gas. I paid the bill with my credit card and had just walked through the front door of the restaurant out into the bright sunlight. I had the receipt from the cashier in my hand and was about to ball it up and toss it away, when I happened to glance at the total to see

UFOS DEJA VU

what I had just spent. The bill – with the tax – came to exactly $20.12! Not a penny more or less!

Someone suggested that the cashier knew me and made out the bill for this amount as a joke. Hardly the case. I had never seen the person before in my life and the receipt was from the checkout register – IT HAD NOT BEEN WRITTEN BY HAND. "Someone" – up there? – saw the humor in slipping me a receipt that came to the identical numerical amount that was on everybody's lips in Sedona. I guess I have had the final laugh, though, as I have told the story on the air now several times.

I would say that the vortex along Airport Mesa was highly influential in the matter of this synchronicity. It can't just be by coincidence that this dollar total would come up on my receipt. Hey, and the Bloody Mary's I drank had nothing to do with this at all, wink, wink, nudge, nudge.

But there is more. "Incident Two," happened right after our visit with APRO librarian Allan Benz. As usual when we got to Sedona, we wandered into the Coffee Pot. After chitchatting over a glass of wine and some Eggs Benedict (with coffee to go naturally), Charla and I wandered out into the parking lot.

I'm not sure who noticed the car parked right next to our vehicle, but it was sticking out a bit more than it should from its parking space. We couldn't believe our eyes. We had to get closer to the car to make sure we weren't hallucinating in the noonday sun. But we have photographs to prove that the license plate on the car parked next to us said, in big bold license plate letters that screamed – APRO.

There is no APRO! The group hasn't existed for over twenty years. So how could there be a license plate on the vehicle parked two feet away that has these four giant capitalized letters like it was a vanity plate smacking us in the face? And even if someone was, say, an old member of the group and still identified with it, why would their car be parked next to ours in the parking lot, or anywhere else in town for that matter? It's more likely you would win the lottery. It's not by "pure chance." It has to be a planned "secret operation."

It doesn't add up, my friends. It's just downright scary, and there is no "rational explanation" that will suffice. So don't accept any of Carl Jung's theories about dreams and deja vu; it has nothing to do with synchronicities based on the assumptions he made for why events repeat themselves.

The vortex is screwing with our heads. Hey, but I'm not paranoid enough to let it trouble me. It just fuels my appetite to learn more about the "chess playing" masterminds from "outside the box" who are attempting to signal to us to get our attention – especially while in Sedona. But why?

I have a lot more questions, but fewer answers. All I know portals are at play in Sedona so you can rest assured if its your "lucky day" something strange is bound to happen.

UFOS DEJA VU

< Though the world did not come to an end, here is proof of Tim's synchronicity in the form of receipt showing that the bill from the Coffee Pot—with the tax — totaled $20.12 exactly — supposedly representing the end of the Mayan calendar in the year 2012. Photo Charla Gene.

After visiting with the former librarian for APRO, an organization that has been defunct for over 20 years, Tim and Charla ended up at the Airport Mesa Vortex in Sedona parked in the Coffee Pot's lot next to a car with this vanity plate. Besides, no one knew they were traveling to Sedona. Photo Charla Gene. >

< The aliens stop whatever they are doing when Beckley arrives in town.

The Coffee Pot is both a restaurant and mother nature's most scenic attraction.

UFOS DEJA VU

ITS NOT A "VICE" TO HAVE A UFO SIGHTING IN SEDONA
By Timothy Green Beckley

PS: And something pretty strange just did a few moments ago while proofing the last chapter. On my Facebook page there was a post that the actress Camille James Harman and I have been FB friends for 10 years. And yet we really have not chatted in some time (2 years maybe?). What are the chances we would be caught up in a conversation a decade later to the precise date that we became social media friends? If you ask me its proof that the pull of Sedona's vortexes can be felt even if you are not there, but hundreds – if not thousands — of miles away.

I had just sent a message to Camille, who I had originally met through Charla Gene while she was still living in Tucson. She had invited Charla and me over for dinner so that I could talk to her about her involvement with the performer, Sting, with whom she once discussed crop circles. Turns out Camille is an abductee with some experiences parallel to those of Whitley Streiber. Her husband, Jeffrey Harman, is a professional astrologer and so we had a great evening, the four of us. Later on, Camille shared with me a photograph of a UFO she had taken hovering alongside Bell Rock which I had published in my Conspiracy Journal newsletter., issue #37. I needed to get some information on the sighting since I had forgotten the details and wanted to publish the photo here as it falls into the context of our subject.

Camille has since moved from the suburbs of Tucson to Hollywood where she has established herself as an actress/producer, and a darn good, accredited one at that.

The winner of 7 achievement awards, Camille is probably best known as Vice President Dick Chaney's secretary in the box office hit, "Vice."

To me the UFO photo Camille took just to the side of Bell Rock looks like one of the space amoeba or "space animals" that Trevor James Constable made famous by using infrared film.

Camille sent me a text on Facebook's Messenger concerning what she remembered about the incident.

"At the time of this photo, I was visiting Sedona with my mother. We were in a car sightseeing with a friend who lived in Sedona, who was also an abductee like I am. We noticed a home with a ladder going into a hole in the rock face and took a photo. We didn't see anything unusual in the sky at the time. Later, when we downloaded the photos, we noticed this strange object just off to the cliff's right hand side." Why are there so many sightings in and around Bell Rock? Camille has a theory: "From what I understand, the mineral content of the mountains of Sedona are conducive to magnetic anomalies."

Unlike others who have felt the pull of the vortexes, Camille now believes the aliens who abducted her should be considered as "demons." I'm not certain why that is, though I know her experience was not necessarily a "welcome one," but that's a topic for a later discussion, and in the meanwhile we wish her continued success on her acting career which I have a hunch may someday lead to an Academy Award.

A POSSIBLE HEALING EFFECT

I thought to balance out any overall negativity the reader might perceive, I would conclude with this positive, beneficial, report posted by an individual who identifies himself only as "AL," but you can verify his existence by traipsing over to UFODigest.Com to see his original comments. AL is the founder of IndigoWorld.com – an interactive forum for Indigo Children, teens and adults. He is a futurist and medium and lectures on metaphysics, personal development and the paranormal. You can contact AL by email at AL@IndigoWorld.com which is the latest information we have. One of his experiences—guess what—pertained to a sighting of an orb over at the Bradshaw Ranch, as we make a 360 degree circle, and arrive back to where we pretty much began our journey in Sedona.

"The energy in the area was one of serenity and even camaraderie between strangers. Some suggest that these orbs have a healing effect. I feel good; happy, as I recall, but I did not come to be healed and this is no Lourdes. Our guide reminded me of the weird phenomena which has occurred here in the recent past and it becomes apparent that we are indeed witness to evidence of an interdimensional portal. Linda Bradshaw herself captured images of a nonexisting oceanside in her photographs there and new reports tell of people hearing 40's music and soda bottles found indicative of that era. This truly is the 'outer limits' of investigation and one must proceed with caution, if at all. Ideas of being transported into some unknown zone seem potentially plausible and sensing and feeling is the order of the day over violating and provoking. I stay, as do we all, behind the barb wire fence and respect the energy, whatever it is, that has chosen to open itself up at the mysterious Bradshaw Ranch."

TOM'S TAKE ON THE VORTEXES

So why all this hubbub about vortexes in and around the Southwestern United States, particularly in Sedona? Tom Dongo says he has a pretty good hunch. "I

UFOS DEJA VU

have come to the conclusion that, along with the world-famous energy vortexes, there are also interdimensional portals in this area. These portals or windows, seem to be entry and exit points into another dimension or universe or place that we, at this point, do not fully understand. This sort of thing isn't happening only in Sedona. Around the world there are a number of these anomalies developing or, perhaps better put, just being discovered. My estimate would be that there are probably hundreds of these portal-type anomalies spread around the globe."

Tom correctly points out that, "UFO and other paranormal activity often is clustered in specific areas, and after a time the attention of researchers is unavoidably and irresistibly drawn to these areas. Crafts and other strange flying objects are seen and photographed far more frequently in these usually small zones. Unknown objects are often seen suddenly appearing or disappearing in and out of these mysterious zones. The logical conclusion is that these anomalies are entry and exit points for these travelers or objects from who knows where. Many of these anomalous zones around the world are located in regions of high magnetic or electromagnetic activity. Sedona, with its red rocks full of iron particles, is certainly one of those magnetic regions."

Tom says that the energy level is different when you are in close proximity to one of these energetic anomalous vortexes. You can begin to feel "strange" and odd things have been known to happen.

Just what is Tom talking about?

"Strange things happen in these portal areas in a variety of ways. Back in 1994, during a one-week period, I shot five rolls of film, a total of 120 photos. I used two cameras, and both were in excellent condition. Only about 30 of these photos developed. All were nighttime shots taken either with flash or five-minute open-shutter timed exposures. Some rolls were lost by developers, and in one case, frames were completely missing from a roll of negatives. Some of these shots I knew had something unusual on them. I had sent these particular rolls to three different out-of-town photo labs, so the finger can't be pointed toward one source. It's also odd that even under "normal" conditions about 80 of my nighttime flash shots developed completely black.

"We all have had batteries in cameras or camcorders suddenly go dead while taking photos in these portal areas. Just days ago (May 1995) I had my Minolta 35mm SLR set up on a tripod to take time-exposure night shots. After several exposures the alkaline batteries suddenly went dead, even though my camera has a low-battery warning light. The warning light never came on. This has now happened twice. I carry a battery tester, and one of the two batteries was totally discharged. I often wonder if an alien life form is doing its level best to communicate with and exchange with us these portal areas, and it, or they, are doing it in the only way they know how, resulting in paradoxes and mysteries creating confu-

UFOS DEJA VU

sion in our human zone of reality."

Whatever the circumstances you find yourself in, being in the red rocks of Sedona may provide you with the opportunity to experience something "special." And even if you don't by chance pick up on the vibe of one of vortexes in town, you can still have a more than decent time. And, hey, send me a couple of snap shots of your trip and your experiences and I will publish them with your permission somewhere along this – or another – time line.

WANT TO LEARN MORE?

Oh, by the way, if you want to learn more about Camille's experiences check out the YouTube episode, "The Musician Sting, Crop Circles and Camille James Harman UFO Abductee" – ://www.youtube.com/watch?v=xwEEAfVoZzs

Actress Camille James Harman appeared as Vice President Dick Chaney's secretary in the movie "Vice." She is a UFO abductee who once discussed the subject of crop circles with rocker Sting.

Camille unknowingly captured this image of a UFO near Bell Rock in 2009.

Pilots approaching the Sedona airport claim they can sometimes feel the pull of the nearby vortex. Don't worry it's not the Bermuda Triangle!

SECTION THREE
TRUE STORIES OF SHAPESHIFTERS, SKINWALKERS AND FLYING OBJECTS

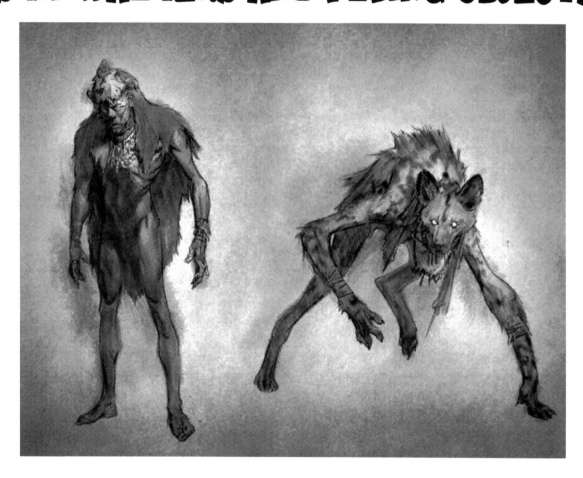

UFOS DEJA VU

UFOS DEJA VU

UFO HISTORY IN THE UINTAH BASIN
By Timothy Green Beckley

Sightings around the Uintah Basin are nothing new.

A search through historical records shows that sometime around 1770 Spanish explorers passing through the area sat around a campfire at night and reported seeing a craft hovering in the sky over what today is the site of the Sherman Ranch.

One local man who has done odd jobs around the ranch and resided within miles had been collecting reports — 400 of them — dating back to the 1950s in the Basin. Joseph "Junior" Hicks (now deceased) was a middle school teacher who had befriended the local Native Americans, many of whose children went to the school where he was an instructor. The unidentified aircraft ranged in size from 20 feet to the size of a football field, and included discs, fireballs, cigars and boomerangs. Several observers said they saw some sort of "living being" inside the craft, through portholes or windows. A percentage of the UFOs were surrounded by glowing lights, others shot down beams onto the ground. The sightings caused a stir in the 1970s, as the Highway Patrol began to receive so many reports that they stopped keeping records of the sightings.

"I'd estimate over 10 percent of the population of the Uintah Basin has seen something," Hicks told the Desert News. "I think what's happening is we are being visited from beings from another world or some other place. I think primarily it's research and exploration."

It was through happenstance, in 1966, that Dr. Frank H. Salisbury encountered Mr. Hicks. "I was giving a lecture on the possibility of life on Mars in the Roosevelt High School, and after my talk Junior Hicks introduced himself to me and told me that he had collected quite a number of UFO reports from the Basin where he had spoken to the witnesses and was convinced they were telling the truth. I was excited by his revelations and told him I wanted to accompany him on some of his field investigations." Hicks agreed and so began a partnership of researcher and scientist. In January 1967, Salisbury published "The Scientist and the UFO" in Bioscience, and the publisher asked him to expand it into a book. Thus began his

UFOS DEJA VU

investigation of UFO reports in the Uintah Basin with the assistance of Hicks, eventually publishing "The Utah UFO Display: A Biologist's Report" in 1974, which has recently been revised.

Born in Utah in 1926, Frank Boyer Salisbury graduated from the University of Utah with a B.S. (1951) and M.A. (1952). He received his Ph.D. from the California Institute of Technology (1955). He was an assistant professor of botany at Pomona College from 1954-1955, then assistant professor and professor of plant physiology at Colorado State University from 1955-1966. He served as a professor and head of the Department of Plant Science at Utah State University from 1966 to 1970, when he resigned as department head to devote more time to researching UFOs.

On an episode of the Paracast.com aired shortly before his death, Dr. Salisbury admitted to hosts Gene Steinberg and Christopher O'Brien that while he felt certain that literally hundreds of sightings of strange craft had been made within short distances of the Sherman Ranch he had never been able to pin down any of the strange events which later had been said to have happened on the ranch itself. "I knew the previous owner, a Mr. Meyers, who had owned the ranch before selling it to Terry Sherman, and he told me he had never encountered anything unusual to speak of." It took him a while to become convinced something "pretty peculiar" was transpiring there, though it did not prevent him from interviewing those local residents who had come face-to-face with something mighty damn weird.

The following are what he admits are the barest skeletons of seven cases taken from Junior Hicks' files from the 1980s, plus one reported to Dr. Salisbury himself.

1. Dean Powell was leaning on the back of his mail truck around noon when he saw a silvery object hovering in front of a bluff less than a quarter mile away. He was joined by another witness who saw the object before it departed rapidly eastward.

2. On 26 September 1966, Kent Denver and friends watched a red ball of fire dance around on the horizon for about 30 minutes.

A little later in the evening, Verl and Leah Haslem were driving home from a bank party when they saw a brilliant red object hovering above their house. As they sped toward home about a mile away, the object moved off to the left (southeast) and departed into the sky "like a meteor in reverse."

At almost exactly the same time, Joe Ann Harris, with four Indian girls in the back seat of her car and a large Indian woman in the front seat, encountered a UFO. The object first appeared as a flashing light at a distance but approached their automobile rapidly until it was 50 to 100 feet in front of the car. It was at least 50 feet in diameter, flat on the bottom with a dome on top. The Indian girls were terrified and on the floor in the back of the car.

The large Indian woman was under the glove compartment. Joe Ann was backing her car away from the object when another automobile approached from the rear. The object apparently left during an instant while Joe Ann had her head turned, backing the car. At about the same time, Estel Manwaring was driving 2 or 3 miles away with an Indian girl in the front seat. They apparently saw the same object, this time with brilliant lights flashing around the rim. After parking to get a better look, the object went straight up with extreme velocity.

3. Thyrena Daniels encountered after dark a huge spherical object, flat on the bottom, glowing red, with bluish to reddish "flames" spouting out horizontally from each side. The UFO, remaining within 100 to 200 feet, preceded her while she drove nearly 30 miles from Vernal to Roosevelt, Utah.

4. A brilliant red, glowing UFO moved across the Uintah Basin while being observed by at least 40 scattered witnesses (several in groups). Richard Hackford saw the object approaching while driving a truck up a hairpin-turn logging road in the Uintah Mountains. The object hovered above the cab of his truck, brilliantly illuminating the surroundings. Night-shift workers at the phosphate plant, perhaps 20 miles away, saw the object move across the basin, hover for a moment where Hackford was driving his truck, and then go into the sky, again "like a meteor in reverse."

5. The Clyde McDonald children told their parents that an Oija Board informed them that a UFO would appear at 8:00 P.M. above the Roosevelt Hospital. The children went out to meet it, and indeed the glowing red sphere was there just as they had predicted! Their parents and several other people in the town were also witnesses.

6. Curtis, Kevin, and Bevin Ercanbrack (brothers) were driving a tractor out to a field to bury a dead calf when, in broad daylight, they saw a large silver object hovering in the field. Before the object left, the three boys came within 50 to 100 feet of it, according to later estimates.

7. Morlin Buchanan and Richard Faucett were hunting geese on Pelican Lake at about sunset when they observed what appeared to be a perfectly spherical balloon with a "string" hanging from below. After driving 8 miles with stops to observe the object through binoculars along the way, they came within perhaps a quarter mile of it, describing it as a huge sphere lit up on the top and bottom (by then it was dark) and with a column of spinning dust or something hanging from beneath. The column was about 4 feet in diameter (referred to by them as a "wind tunnel") and perfectly even in diameter top to bottom (otherwise reminding them of a tornado spout). Finally, the "wind tunnel" seemed to be sucked rapidly into the object, which then disappeared with a flash of light.

8. Leland Mecham and his young son, Jody, were riding home on horseback late at night when they saw a huge, glowing red object perhaps 5 to 7 miles away, almost on the horizon but in front of more distant mountains. Beams of light seemed to be emanating toward the ground from the object, and clouds of billowing red

UFOS DEJA VU

"dust" could be seen where the rays struck the ground. An arch of white light extended above the object. While Leland and Jody lost sight of the object behind some trees, Dee Hullinger and his wife watched it spin once on its axis and shoot into the sky with high velocity. The red glow and the white "dome" of light remained for an hour or so, gradually becoming dimmer until they had disappeared.

Those wishing to explore in depth the history of the sightings in the Basin around the ranch are invited to read "The Utah UFO Display: A Scientist Brings Reason and Logic to over 400 Sightings in Utah's Uintah Basin."

Joe "Jr" Hicks and Dr. Frank Salisbury
consult on sightings in Utah.

Above left:
The recently updated book that spilled the
beans on the history of UFOs in Utah.

UFOS DEJA VU

The 3 Amigo's — Dr. Jacques Vallee, Dr. Frank Salisbury and Junior Hicks. Courtesy PhantomHero Radio.

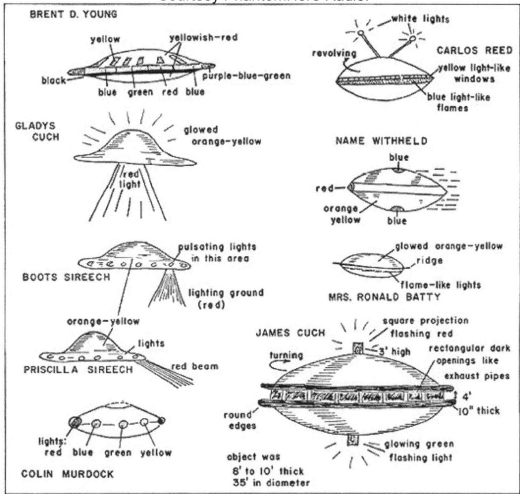

Fig. 3. Other UFO drawings as depicted by witnesses. The originals were done with colored pencils on the bottoms of the reports. A photograph was made, and then a tracing from the photograph. (From Salisbury 1974 with permission.)

UFOS DEJA VU

Portals and Passages! Vortexes and Merging Dimensions! If you can wrangle a pass to get inside the Bradshaw Ranch, you may suddenly encounter portals, shapeshifters, skinwalkers, ETs and MORE, as in this artist's conception.

UFOS DEJA VU

SKINWALKERS AND SHAPESHIFTERS
THE WHO, WHAT, AND WHERE!
By The Whistle Blower Known Simply As "Think About It"

PUBLISHER'S NOTE: In order to do a complete study of the vast variety of unexplained phenomena taking place, not only on the Sherman Ranch and the Uintah Basin, but around many of the other portals located in the Four Corners, one has to take into consideration the legends and lore of the skinwalkers and shapeshifters. These legends persist among the various American Indian tribes of the great southwestern USA. And who better to do this than the unanimous "Think About It" who for over 20 years has been trying to help people think for themselves, to learn that not everything you're told is the truth. This site began when there was not a lot of alternative information out there, and finding it took many hours, days or months, sometimes years, to locate only the smallest little bits. This was the time when just a few outspoken persons, had access to this information which could also mean trouble. Big Trouble! "Think About It" was one of the first to speak about the underground bases, aliens, alternative spirituality, UFOs and other weird things. "We shared the Branton Papers, works from William Cooper, Bill Hamilton, and many other 'truth walkers' with the public," and don't you dare forget about it! And, yes, there are more recent rantings about all of this on ThinkAboutIt.com

What role does Navajo witchcraft play in what some see as a curse placed on Skinwalker Ranch, and ultimately being responsible for all the strange activity going on there? For a possible answer we must study the principalities of the skinwalker, a big part of the oral stories told by Native American culture, usually in hushed tones and whispers so that they may not ultimately be cursed and have to fend off these half-humans, half-creatures.

With skinwalkers becoming the subjects of popular books and recently, movies, thanks to journalist George Knapp and a handful of researchers and academics, it is fair to ask about their origins. In August 1996, a team of scientists arrived on a remote ranch in Northeast Utah to investigate a bizarre litany of phenomena; including unidentified flying objects, animal mutilations, paranormal and poltergeist occurrences that appeared to erupt almost on a nightly basis. The list went on and on. The first piece of information the team learned from local people was that the ranch lay "on the path of the skinwalker." Was the skinwalker responsible for the weird happenings on this ranch? What followed was a multitiered odyssey into the dark unknown as the science team tried to pursue, measure and photograph the elusive skinwalker. The complete account of the unprecedented research project is published in the book "Hunt For The Skinwalker," by the popular weekend host of "Coast To Coast AM."

In the religion and cultural lore of Southwestern tribes, there are witches known as skinwalkers who can alter their shapes at will to assume the characteristics of certain animals. Most of the world's cultures have their own shapeshifter legends. The best known is the werewolf, popularized by dozens of Hollywood movies. European legends as far back as the 1500's tell stories about werewolves. (The modern psychiatric term for humans who believe they are wolves is lycanthropy.) The people of India have a were-tiger legend. Africans have stories of were-leopards and were- jackals. Egyptians tell of were-hyenas.

In the American Southwest, the Navajo, Hopi, Utes, and other tribes each have their own version of the skinwalker story, but basically they boil down to the same thing–a malevolent witch capable of transforming itself into a wolf, coyote, bear, bird, or any other animal. The witch might wear the hide or skin of the animal identity it wants to assume, and when the transformation is complete, the human witch inherits the speed, strength, or cunning of the animal whose shape it has taken.

"The Navajo skinwalkers use mind control to make their victims do things to hurt themselves and even end their lives," writes Doug Hickman, a New Mexico educator. "The skinwalker is a very powerful witch. They can run faster than a car and can jump mesa cliffs without any effort at all."

* * * * *

MORE THAN JUST A LEGEND

For the Navajo and other tribes of the southwest, the tales of skinwalkers are not mere legend. Just ask Michael Stuhff. A Nevada attorney, Stuhff is likely one of the few lawyers in the history of American jurisprudence to file legal papers against a Navajo witch. He has often represented Native Americans in his practice. He understands Indian law and has earned the trust of his Native American clients, in large part because he knows and respects tribal religious beliefs.

As a young attorney in the mid-70s, Stuhff worked in a legal aid program based near Genado, Arizona. Many, if not most, of his clients were Navajo. His legal confrontation with a witch occurred in a dispute over child custody and financial support. His client, a Navajo woman who lived on the reservation with her son, was asking for full custody rights and back child support payments from her estranged husband, an Apache man. At one point during the legal wrangling, the husband got permission to take the son out for an evening, but didn't return the boy until the next day. The son later told his mother what had transpired that night.

According to the son, he spent the night with his father and a "medicine man." They built a fire atop a cliff and, for many hours, the medicine man performed ceremonies, songs, and incantations around the fire. As dawn broke, the three traveled into a wooded area near a cemetery, where they dug a hole. Into the hole, the medicine man deposited two dolls made of wood. One of the dolls was made of dark wood, the other of light wood. It was as if the two dolls were meant to represent the mother and her lawyer. Although Stuhff wasn't sure how seriously to take the news, he recognized that it certainly didn't sound good, so he sought out the advice of a Navajo professor at a nearby community college.

"He told me that the ceremony I had described was very powerful and very serious and that it meant that I was supposed to end up buried in that cemetery," Stuhff says. "He also said that a witch can perform this type of ceremony only four times in his life, because if he tries it more than that, the curse would come back on the witch himself. He also told me that if the intended victim found out about it, then the curse would come back onto the person who had requested it."

Stuhff thought about a way to let the husband know that he had found out about the ceremony, so he filed court papers that requested an injunction against the husband and the unknown medicine man, whom he described in the court documents as "John Doe, A Witch." The motion described in great detail the alleged ceremony. The opposing attorney appeared extremely upset by the motion, as did the husband and the presiding judge. The opposing lawyer argued to the court that the medicine man had performed "a blessing way ceremony," not a curse. But Stuhff knew that the judge, who was a Navajo, could distinguish between a blessing ceremony, which takes place in Navajo hogans (homes), and what was obviously a darker ceremony involving lookalike dolls that took place in the woods near a cemetery. The judge nodded in agreement when Stuhff responded. Before the judge could rule, Stuhff requested a recess so that the significance of his legal motion could sink in. The next day, the husband capitulated by agreeing to grant total custody to the mother and to pay all back child support.

"I took it very seriously because he took it seriously," Stuhff says. "I learned early on that sometimes witches will do things themselves to assist the supernatural, and I knew what that might mean."

Whether or not Stuhff literally believes that witches have supernatural powers, he acknowledges that this belief is strongly held in the Navajo nation. Certain communities on the reservation had reputations as witchcraft strongholds, he says. It is also unknown whether the witch he faced was a skinwalker or not. "Not all witches are skinwalkers," he says, "but all skinwalkers are witches. And skinwalkers are at the top. They are a witch's witch, so to speak."

According to University of Nevada Las Vegas anthropologist Dan Benyshek, who specializes in the study of Native Americans of the Southwest, "Skinwalkers are purely evil in intent. I'm no expert on it, but the general view is that skinwalkers do all sorts of terrible things—they make people sick, they commit murders. They are grave robbers and necrophiliacs. They are greedy and evil people who must kill a sibling or other relative to be initiated as a skinwalker. They supposedly can turn into were-animals and can travel in supernatural ways."

Benyshek and other scientists do not necessarily endorse the legitimacy of the legends, but they recognize the importance of studying stories about skinwalkers because the power of the belief among Native Americans manifests itself in ways that are very real. "Oh, absolutely," Benyshek explains. "Anthropologists have conducted scientific investigations into the beliefs in Native American witchcraft because of the effects of such beliefs on human health."

Anthropologist David Zimmerman of the Navajo Nation Historic Preservation Department explains: "Skinwalkers are folks that possess knowledge of medicine, medicine both practical (heal the sick) and spiritual (maintain harmony), and they are both wrapped together in ways that are nearly impossible to untangle."

THE DARK SIDE

As Zimmerman suggests, the flip side of the skinwalker coin is the power of tribal medicine men. Among the Navajo, for instance, medicine men train over a period of many years to become full fledged practitioners in the mystical rituals of the Dine' (Navajo) people. The U.S Public Health Service now works side by side with Navajo medicine men, because the results of this collaboration have been proven, time and again, in clinical studies. The medicine men have shown themselves to be effective in treating a range of ailments.

"There has been a lot of serious research into medicine men and traditional healers," says Benyshek. "As healers, they are regarded as being very effective in some areas."

But there is a dark side to the learning of the medicine men. Witches follow some of the same training and obtain similar knowledge as their more benevolent colleagues, but they supplement both with their pursuit of the dark arts, or black magic. By Navajo law, a known witch has forfeited its status as a human and can be killed at will. The assumption is that a witch, by definition, is evil.

"Witchcraft was always an accepted, if not widely acknowledged part of Navajo culture," wrote journalist A. Lynn Allison. "And the killing of witches was historically as much accepted among the Navajo as among the Europeans." Allison has studied what she calls the "Navajo Witch Purge of 1878" and has written a book on the subject. In that year, more than 40 Navajo witches were killed or

UFOS DEJA VU

"purged" by tribe members because the Navajo had endured a horrendous forced march at the hands of the U.S. Army in which hundreds were starved, murdered, or left to die. At the end of the march, the Navajo were confined to a bleak reservation that left them destitute and starving. The gross injustice of their situation led them to conclude that witches might be responsible, so they purged their ranks of suspected witches as a means of restoring harmony and balance. Tribe members reportedly found a collection of witch artifacts wrapped in a copy of the Treaty of 1868 and "buried in the belly of a dead person." It was all the proof they needed to unleash their deadly purge.

"Unexplained sickness or death of tribal members or their livestock could arouse suspicion of witchcraft," wrote Allison in her book. "So could an unexplained reversal of fortune, good or bad."

Read: "2016 After Years Of Oppression U.S. To Pay Out Nearly $1 Billion To Native Americans".

The Navajo people do not openly talk about skinwalkers, certainly not to outsiders. Author Tony Hillerman, who has lived for many years among the Navajo, used the skinwalker legend as the backdrop for one of his immensely popular detective novels, one that pitted his intrepid Navajo lawmen Jim Chee and Joe Leaphorn against the dark powers of witchcraft. The following excerpt is from Skinwalkers:

"You think that if I confess that I witched your baby, then the baby will get well and pretty soon I will die," Chee said. "Is that right? Or, if you kill me, then the witching will go away." It was agreed! "You should confess," the woman said. "You should say you did it. Otherwise, I will kill you."

Hillerman has been harshly criticized by some Navajo for bringing unwanted attention to the subject of skinwalkers. "No one who has ever lived in the Navajo country would ever make light of this sinister situation," wrote one critic after Hillerman's book was produced as a movie that aired on PBS in 2003.

Anthropologist Zimmerman explains why so little information is available on skinwalkers: "Part of the reason you won't find a lot of information about skinwalkers in the literature is because it is a sensitive topic among the Dine'. This is often referred to as proprietary information, meaning it belongs to the Dine' people and is not to be shared with the non-Dine'"

We know from personal experience that is it extremely difficult to get Native Americans to discuss skinwalkers, even in the most general terms. Practitioners of adishgash, or witchcraft, are considered to be a very real presence in the Navajo world. Few Navajo want to cross paths with naagloshii (or yee naaldooshi), otherwise known as a skinwalker. The cautious Navajo will not speak openly about skinwalkers, especially with strangers, because to do so might invite the attention of an evil witch. After all, a stranger who asks questions about skinwalkers

UFOS DEJA VU

just might be one himself, looking for his next victim.

THEIR EYES GLOW RED

"They curse people and cause great suffering and death," one Navajo writer explained. "At night, their eyes glow red like hot coals. It is said that if you see the face of a Naagloshii, they have to kill you. If you see one and know who it is, they will die. If you see them and you don't know them, they have to kill you to keep you from finding out who they are. They use a mixture that some call corpse powder, which they blow into your face. Your tongue turns black and you go into convulsions and you eventually die. They are known to use evil spirits in their ceremonies. The Dine' have learned ways to protect themselves against this evil and one has to always be on guard."

One story told on the Navajo reservation in Arizona concerns a woman who delivered newspapers in the early morning hours. She claims that, during her rounds, she heard a scratching on the passenger door of her vehicle. Her baby was in the car seat next to her. The door flung open and she saw the horrifying form of a creature she described as half-man, half-beast, with glowing red eyes and a gnarly arm that was reaching for her child. She fought it off, managed to pull the door closed, then pounded the gas pedal and sped off. To her horror, she says, the creature ran along with the car and continued to try to open the door. It stayed with her until she screeched up to an all-night convenience store. She ran inside, screaming and hysterical, but when the store employee dashed outside, the being had vanished. Outsiders may view the story skeptically, and any number of alternative explanations might be suggested, but it is taken seriously on the Navajo reservation.

Although skinwalkers are generally believed to prey only on Native Americans, there are recent reports from Anglos claiming they had encountered skinwalkers while driving on or near tribal lands. One New Mexico Highway Patrol officer revealed that while patrolling a stretch of highway south of Gallup, New Mexico, he had two separate encounters with a ghastly creature that seemingly attached itself to the door of his vehicle. During the first encounter, the veteran law enforcement officer said the unearthly being appeared to be wearing a ghostly mask as it kept pace with his patrol car. To his horror, he realized that the ghoulish specter wasn't attached to his door after all. Instead, he said, it was running alongside his vehicle as he cruised down the highway at a high rate of speed.

PUBLISHER'S NOTE: For what it is worth – we have several UFO cases in our files where the Ultra-terrestrials chased or kept pace with an automobile's driver, running down the road beside the car door at speeds of 60 or so miles per hour. So when does a UFOnaut become a specter?

OFFICIAL POLICE REPORTS

The officer said he had a nearly identical experience in the same area a few days later. He was shaken to his core by these encounters, but didn't realize that he would soon get some confirmation that what he had seen was real. While having coffee with a fellow highway patrolman not long after the second incident, the cop cautiously described his twin experiences. To his amazement, the second officer admitted having his own encounter with a white-masked ghoul, a being that appeared out of nowhere and then somehow kept pace with his cruiser as he sped across the desert. The first officer told us that he still patrols the same stretch of highway and that he is petrified every time he enters the area.

One Caucasian family still speaks in hushed tones about its encounter with a skinwalker, even though it happened in 1983. While driving at night along Route 163 through the massive Navajo Reservation, the four members of the family felt that someone was following them. As their truck slowed down to round a sharp bend, the atmosphere changed, and time itself seemed to slow down. Then something leaped out of a roadside ditch at the vehicle.

"It was black and hairy and was eye level with the cab," one of the witnesses recalled. "Whatever this thing was, it wore a man's clothes. It had on a white and blue checked shirt and long pants. Its arms were raised over its head, almost touching the top of the cab. It looked like a hairy man or a hairy animal in man's clothing, but it didn't look like an ape or anything like that. Its eyes were yellow and its mouth was open."

The father, described as a fearless man who had served two tours in Vietnam, turned completely white, the blood drained from his face. The hair on his neck and arms stood straight up, like a cat under duress, and noticeable goose bumps erupted from his skin. Although time seemed frozen during this bizarre interlude, the truck continued on its way, and the family was soon miles down the highway.

A few days later, at their home in Flagstaff, the family awoke to the sounds of loud drumming. As they peered out their windows, they saw the dark forms of three "men" outside their fence. The shadowy beings tried to climb the fence to enter the yard but seemed inexplicably unable to cross onto the property. Frustrated by their failed entry, the men began to chant in the darkness as the terrified family huddled inside the house.

The story leaves several questions unanswered. If the beings were skinwalkers, and if skinwalkers can assume animal form or even fly, it isn't clear why they couldn't scale a fence. It is also not known whether the family called the police about the attempted intrusion by strangers.

The daughter, Frances, says she contacted a friend, a Navajo woman who is knowledgeable about witchcraft. The woman visited the home, inspected the grounds, and offered her opinion that the intruders had been skinwalkers who were drawn by the family's "power" and that they had intended to take that power by whatever means necessary. She surmised that the intrusion failed because something was protecting the family, while admitting that it was all highly unusual, since skinwalkers rarely bother non-Indians. The Navajo woman performed a blessing ceremony at the home. Whether the ceremony had any legitimacy or not, the family felt better for it and has had no similar experiences in the ensuing years.

This disturbing account is not offered as definitive proof of anything, particularly since we have not personally interviewed the witnesses. It is presented only as an illustration of the intense fear and unsettling descriptions that permeate skinwalker lore, and which are accepted at face value by the Native Americans for whom the skinwalker topic is not just a spooky children's story.

The ranch property has been declared as off-limits to tribal members because it lies in the path of the skinwalker. Even today, Utes refuse to set foot on what they see as accursed land. But the tribe doesn't necessarily believe that the skinwalker lives on the ranch. Hicks says the Utes told him that the skinwalker lives in a place called Dark Canyon which is not far from the ranch. In the early 1980's, Hicks sought permission from tribal elders to explore the canyon. He's been told there are centuries-old petroglyphs in Dark Canyon, some of which depict the skinwalker. But the tribal council denied his request to explore the canyon. One member later confided to Hicks that the tribe denied the request, because it did not want to disturb the skinwalker for fear that it might "create problems." The tribe's advice to Hicks: "Leave it alone."

READ: The Bear Warrior: A Native American Myth

Dan Banyshek suggests that some parts of this account don't add up. He thinks it unlikely that the Navajo would enlist the assistance of a skinwalker to carry out their revenge on the Utes, no matter how much the tribe might want some payback on their enemy. "The skinwalkers are regarded as selfish, greedy, and untrustworthy," Banyshek says. "If the Navajo knew someone to be a skinwalker, they would probably kill him, not ask for his help with the Utes. Besides, even if he was asked, the skinwalker would be unlikely to help the Navajo get revenge, since his motives are entirely evil and self-serving. From the Navajo perspective, this story doesn't make sense."

THE CREATURES

But from the Ute perspective, it could ring true. "The Utes could very likely have concluded that the curse is real," explains Banyshek. "Different tribes or bands would often tell stories about the evil motives of other tribes they were in conflict with, about how another tribe was in league with witches, or how other tribes were cannibals. The Utes might tell themselves this story as a way to explain their own misfortunes."

Hicks told us that the Indians say they see them a lot. "When they go out camping," he says, "they sprinkle bark around their campsites and light it as protection against these things. But it's not just Indians. Whites see them, too." Like his Ute neighbors, Hicks sometimes uses the terms skinwalker and Sasquatch interchangeably. He says he's seen photographs of the telltale huge footprints often associated with Bigfoot, taken in the vicinity of the Sherman ranch. But whether it was a run-of-the-mill Sasquatch or a far more sinister skinwalker isn't always clear, even to those who accept the existence of both.

"There was an incident 16 years ago, where a skinwalker was on a porch in Fort Duchesne," Hicks remembers. "They called the tribal police and tracked it east toward the river. They took some shots at it and thought they hit, it because they found blood on the ground, but they never found a body."

We also conducted an interview with a Ute man who worked as a security officer for the tribe. He provided us with details about his own encounter with a Bigfoot or skinwalker. Brandon Ware (not his real name) received his police training at an academy associated with the Bureau of Indian Affairs. He says he was working the 10:00 P.M. to 4:00 A.M. shift, guarding a tribal building near a part of the reservation known as Little Chicago. Between midnight and 1:00 in the morning, Ware walked up to check on the building and noticed that the guard dogs inside were calm but intently staring through a window at something outside. They weren't barking, he said, just looking.

"I could see this big ol' round thing, you know, in the patio over there," Ware recalls, "and the hair started raising on my neck and I kinda got worried a little bit trying to figure out what those things were. I stood there and watched it for a few minutes, then it came over the top and headed down the road. But I could smell it. Even after it was gone, you could smell it."

Ware says that when the creature realized it was being observed, it briefly looked over at Ware, then vaulted over a short wall that surrounded the patio area outside the building. He says it took off running toward the Little Chicago neighborhood, crashing into garbage cans as it moved past the homes, and generating a cacophony of loud barking by every dog in the immediate area. Ware says he then went into the building and telephoned another on-duty officer who was nearby. By the time Ware left the building, the other officer had pulled up in his patrol car.

Ware told the other officer to turn off his engine, so they could listen to the hubbub that was still unfolding among the nearby homes. "We listened a little bit and we could hear it. Then we jumped in and took off. We headed down the hill to see if we could catch up to it."

The two officers didn't see the creature again that night, but had no trouble tracing its path through the cluster of homes, because they were able to follow a noticeable trail of scattered garbage cans. "It must have gone straight on through," Ware recalls. "We could see where cans—people usually tie up their cans—they were all off. I told the other officer, 'hey man, maybe it picked up them cans and was throwing them at those dogs'."

Ware provided us with further details about what he had seen. His initial impression was of something dark and round. But he says that when the creature stood erect to vault over the patio wall, it appeared to be "huge." Ware was carrying a large flashlight at the time of the encounter. He says he was using the flashlight just minutes before the encounter, while checking the doors of the building, but when he tried to use it to illuminate the creature, the light wouldn't turn on. When the creature took off running down the hill, the flashlight clicked back on.

"He moved quick," he told us. "Whatever it was, it moved—I called him a 'he'—it could have been a she. It could have been whatever, but he moved quick going down through there. But it was kind of cool. It was neat. I never knew it….it was something I've never seen before. I've heard about them. I heard the old people talking about some of these things."

A FEW NIGHTS LATER

Just a few nights later, Ware got a chance for a second look. He and another officer, "Bob," were patrolling a back road that emerges at a spot known as Shorty's Hill. They emerged from the road to a pasture area that was punctuated by a large rock. "I don't know if it was the same guy or not," Ware says. "It was a big ol' black hairy thing hanging there, and when it turned around, it had big ol' eyes on him about yea big. We'd just passed it and I told Bob 'there he is,' and then he come to a screeching halt and we backed up. By the time we got out, it was gone."

Ware described the creature's eyes as being "coal red" and unusually large. He isn't sure whether the headlights of the patrol car might have affected his perception of the beast's eye color, but tends to doubt it. He has no doubt about the presence of the beast itself. "We got out there to go look and we had shotguns and pistols and everything. We were going to blow him away," Ware admits.

When pressed for his opinion of what he had seen, whether it might have been a Sasquatch or even a skinwalker, Ware's response seemed to draw a distinction between the two, but the distinction became blurry as the conversation progressed and Ware explained his understanding of tribal lore.

"Sasquatch, he's an old man, an old man that lives on a mountain," he explained. "He just comes in and looks at people and then he goes back out again. He just lives there all his life, never takes care of himself, and just smells real bad. Almost like, almost like that guy, like he is dirty, dirty human being smell is what it smelled like…a real deep, bad odor….It smelled like dirty bad underarms…The closer I got, the worse the smell got." Could the creature he saw have been a skinwalker?

"Nope," said Ware. "A skinwalker's smaller. A skinwalker is the size of humans, six foot and under. They don't come in most of the time to where the animals are at. They come in where people are at. They can come right here and you'd never know he was standing here looking at you in the middle of the night…they can take the shape of anything they want to take the shape of. Like I said, they're medicine."

Ware said that skinwalker sightings among the Utes are not uncommon. He told us of an encounter with two shapeshifters near the Sherman ranch. The figures he described are so unusual, so far outside our own concept of reality as to be almost comical, like something out of a Saturday morning cartoon. One local who saw them on the road in Fort Duchesne described them as humans with dog heads smoking cigarettes. But Ware was perfectly serious in his description. He certainly did not bare his soul for comic effect, and we have no interest in making light of his story. For him, and for many others, skinwalkers are as real as the morning sun or the evening moon. They are a part of everyday life, and they most certainly are integral to the story of the Sherman ranch.

Could the Utes have used the skinwalker curse as an all-encompassing explanation for their assorted tribal misfortunes, as Banyshek asks? Or are they relying on the legend as an umbrella explanation for the wide range of paranormal events that have been reported in the vicinity of their lands for generations—in particular, in the vicinity of the ranch?

If a skinwalker really is a shapeshifter, capable of mind control and other trickery, might it also have the ability to conjure up nightmarish visions of Bigfoot or UFOs? Could it steal and mutilate cattle, incinerate dogs, generate images of monsters , unknown creatures, or extinct species, and could it also frighten hapless residents with poltergeist-like activity?

At the very least, the skinwalker legend might be a convenient way for the Utes to grasp a vast menu of otherwise inexplicable events, the same sort of events that might stymie and confuse a team of modern scientists.

One thing is sure, it is obvious that the legend of skinwalkers has entered popular culture in ways not seen before.

UFOS DEJA VU

Лехорадка. Худ. Надежда Антяхова

Native Americans at tribal gathering indicating that they could turn into a skinwalker at any moment.

Above: Shapeshifters often take on the form of hideous flying creatures.
Balty Mitologija

Above: This posting of a shapeshifter is being passed off as being authentic. What do you believe?

Right: This official Canadian postage stamp would seem to indicate a belief in this creature seen on many Native reservations.

CANADA

THE SASQUATCH
LE SASQUATCH

UFOS DEJA VU

A morphing monster off of a YouTube channel.

Shapeshifters are known to exist all over the world as portrayed in Slavic painting circa 1906.

Supposed skinwalkers in an early photo. The Native Americans who practiced shapeshifting were considered "dark witches."

UFOS DEJA VU

YOU'RE UNWELCOME AT SKINWALKER RANCH
By Sean Casteel

PUBLISHER'S NOTE: Though it's just conjecture— we don't have a CNN poll to go by—I would have to say that, outside of Area 51, no other UFO hot spot has drawn as much attention to itself nor is as well-known as the Skinwalker Ranch.

Regardless of if you reside in New York City or in Timbuktu, and watch any amount of cable TV or listen to late night radio ("Coast to Coast AM" in particular), you will probably have heard of this parcel of ground, reputed to be haunted by all sorts of paranormal and UFO-related activity. Located in the Uintah Basin, the Skinwalker Ranch—also known as Sherman Ranch, named after its last private owner—is a 500+ acre tract of land southeast of the city of Ballard, reputed to be the site of a diversity of paranormal and UFO-related activities. The ranch takes its name from the "skinwalker" of Navajo legend, of shape-shifting, malicious witches who have the supernatural ability to turn themselves into a variety of animals, mainly wolves and flesh-eating birds. By a broader definition, shape-shifting occurs when a being, usually a human, has the ability to change into that of another person, creature, gender, species or other entity.

If you fly into Salt Lake City and rent a car, the 150 mile drive via U.S. 40 East will take about three hours (one stop for gas and a drive through McDonalds) – but here is the kicker: YOU WILL BE UNWELCOME! So you might as well go to Area 51 instead, where at least you can have a drink at the Little A' Le' Inn before turning around at the barrier to the top secret military base made so popular with the "Storm the Gate" movement that took hold – well, it almost did – because of a Facebook posting meant to be a spoof.

So, if you can't score a "special invitation" to the Skinwalker Ranch, we offer the next best thing: Take our escorted tour, which is boarding just about now, with Sean Casteel at the ship's helm.

* * * * *

The history of the legendary Skinwalker Ranch is intertwined with the story of billionaire businessman Robert Bigelow. His interest in the creation of commercially viable spacecraft formed his initial motivation for his later absorption by

UFOS DEJA VU

what are considered "fringe" topics.

From the late 1960s through the 1990s, Bigelow acquired his wealth by developing commercial real estate, to include hotels, motels and apartments. He founded Bigelow Aerospace in 1999. He had long harbored an intense belief, beginning in childhood, that his future lay in space travel. He intended to make enough money to start his own space program. Bigelow kept that ambition secret for many years, even from his wife.

In 1995, Bigelow founded the National Institute for Discovery Science (NIDS) to research and advance the study of UFOlogy as well as investigating cattle mutilations and black triangle reports, ultimately attributing the latter to the military. The institute was disbanded in 2004, though it can hardly be said that he retired from UFO research. Bigelow has indicated that he intends to spend up to $500 million to develop the first commercial space station, and his aerospace company has launched two experimental space modules, called Genesis 1 and Genesis 2. In October 2017, Bigelow announced that he planned to put an inflatable "space hotel" into orbit by 2022.

Also in 2017, Bigelow was reported by the New York Times to have urged Senator Harry Reid to initiate what became the Advanced Aviation Threat Identification Program, a government project which operated from 2007 to 2012 tasked with the study of UFOs. According to the Times, Bigelow had stated publicly that he was "absolutely convinced" that aliens exist and that UFOs have visited Earth. He repeated this claim on the immensely popular "60 Minutes" TV show which goes into millions of homes here in the U.S. and is rebroadcast worldwide.

MORE THAN THEY BARGAINED FOR

 Claims about the ranch Bigelow would come to purchase first appeared in 1996, in the Desert News in Salt Lake City. Investigative journalist George Knapp later wrote a series of articles for an alternative weekly called Las Vegas Mercury. These early stories detailed the claims of the Sherman family, who had allegedly experienced inexplicable and frightening events after they purchased and occupied the property in 1995.

And thus the tale begins, with ranchers Terry and Gwen Sherman, (initially referred to as the Gormans, when they were trying to keep their identity a secret) who could not possibly have foreseen that their large tract of land in Utah would lead them down a paranormal rabbit hole and into another reality. The family found their new ranch unusual from day one, according to UFO researcher Christopher O'Brien, who was one of the first to arrive on the Sherman case. "The house had sat empty for seven years. Any house that sits empty for even a month or two in this area is completely cannibalized to the ground. This place—no one would touch it," says O'Brien, whom we shall bring more fully into the picture in a later chapter about his own involvement in a particular mysterious

UFOS DEJA VU

zone in the state of Colorado.

O'Brien also provided more strange details on the nature of the house, saying it looked like it had been vacated hastily the day before, and all the doors in the house had deadbolt locks. A central corridor could be locked on both ends and a closet in that hallway could be locked from the inside. "It was very spooky – like a Stephen King novel or something," says O'Brien.

In an article for "Spirit Magazine," writer/editor Dave Perkins, who has devoted quite a bit of time to keeping tabs on activity associated with the ranch, and has become our chief source of information on what's going on out there, provides some fascinating details of what the Shermans endured after moving into their "dream ranch."

"Terry Sherman calmly told me their story," Perkins writes. "In the summer of 1994, Terry (a rancher and cattle breeder) and his wife Gwen (employed for 20 years at the local bank,) had found their dream ranch. The spread was a remote little paradise. It would be a fine place, they thought, to raise their teenage son and 9-year-old daughter. They were puzzled why such a prime piece had been sitting vacant for seven years. The land bordered the Uintah and Ouray Ute Indian Reservation and was protected by a long red rock ridge.

"The first signs that something was 'different' about the place were the large, circular impressions which the Shermans kept finding in their pastures. One configuration formed a 30 foot triangle. Other circles were found measuring roughly three feet wide and one to two feet deep. The soil inside the holes was firmly impacted. About this time, Terry began having trouble with his prize breeding herd of cattle. Cows were dying under unexplained circumstances."

In April of 1995, the weirdness dramatically escalated. While checking his cattle one evening, Terry saw a silent glowing object pass over a 50 foot tall stand of poplar trees that fringed one of their fields.

"A few days later, Gwen Sherman saw another unexplained flying object: 'It looked like headlights, but they were a little ways away from the craft. It just lit the whole side of the mountain up like it was broad daylight." Terry started examining his odd cattle deaths more closely. The first cow found dead (shortly after a UFO sighting) showed only a hole in the center of its left eyeball. Predators had not touched the carcass and Sherman noted a chemical smell in the vicinity. A short time later, a second cow was found dead with the same hole in the left eyeball. With both these animals, Terry had taken a wire and inserted it into the hole to gauge its depth. In both cases 'the wire slipped in easily to the center of the brain.'" Also during this time, some of the Shermans' cows started disappearing. As Terry said: 'We contacted everyone around. We looked everywhere. They just vanished.'"

Researcher Dave Perkins spoke to Terry Sherman on and off, taking note of all

UFOS DEJA VU

the strange activity on the ranch, fascinated by what seemed to be the unending stream of claims being made.

"In one instance," Dave (who later went on to establish the very polished online magazine "Shadows of the Mind") writes, "Terry followed the tracks of a cow in fresh snow. The tracks 'just stopped' under some trees at the edge of a field. The area around the animal's last steps was surrounded by a circle of fresh twigs and branches which Terry could see had come from the trees above.

"During the next few months, the Shermans observed a variety of 'craft' and the mutilation activity continued. The most spectacular aerial phenomenon they observed was described by Terry: 'We would see these 100 foot circular openings appear in the air. It was like four orange colored doorways would sort of spiral open.' Looking through a high powered scope, the Shermans watched as smaller craft would emerge from the hovering portals, fly around the property and then reenter the doorways.

The Shermans described the 'stealthy smaller craft' as being about 60 by 40 feet and 'square-ish with short wings.' The smaller craft looked like 'they were flying a grid.' They also appeared to emit 'spikes of light which hit the ground.' The Shermans thought this to be some sort of navigational system.

"In a rare occurrence, the Shermans' son found a mutilated cow within five minutes of its death. The young man had seen the 'gentle' Angus eating peacefully and returned moments later to find it dead. The cow's rectum had been 'cored out' with a 6-inch wide hole that was 8 inches deep.

"During that summer, Terry, his son and his nephew had heard unintelligible voices while standing in a nearby pasture. The sound, which they first assumed to be the echoes of a CB radio, seemed to emanate out of the air directly above them. As they listened more closely, they could distinctly hear two voices speaking in an unknown language, which Terry described as 'choppy' and halting, like a cross between Russian and Native American. One voice had a deep resonant tone and the other was higher pitched. Terry yelled into the air: 'We can hear you!' The voices stopped momentarily and then the deeper voice broke into a low rumbling laugh. The conversation then went on as before.

"By the fall of the year, events seemed to be moving toward a climax. Seeing 'the lights' in a field one night, Gwen grabbed her binoculars. Focusing in, she was shocked to see a 'square lighted structure' sitting on the ground. Before the light blinked out, Gwen caught a glimpse of 'a large, heavy set individual' seated in the object. A short time later, the craft appeared again. This time both she and Terry watched through a 60 power spotting scope. They could see a figure standing next to the object. Terry described the 'person' as being 'over seven feet tall and decked out in a totally black uniform and very huge.'

The Shermans noted that the being appeared to have 'a visor or something

UFOS DEJA VU

shiny on its face' because of the way the light glinted from its head area."

BLUE BALLS ATTACK: DOGS TURNED TO BUTTER

According to the normally very down-to-earth Perkins, "Another eerie phenomena soon began to plague the Shermans. The family started noticing 'glowing blue balls' moving around the property. The balls gave off a 'crackling sound,' seemed intelligently controlled and could either hover or move 'unbelievably fast.'

One evening, the Shermans watched as a blue ball approached one of their horses. The light hovered within a foot of the horse's face, spooking it mightily. From a distance of ten feet, Gwen shined a flashlight on the blue globe and it retreated. It then approached Terry as if 'inspecting' him. Terry described it as 'a glass ball about the size of a baseball,' which appeared to contain 'two blue fluids which intermingled with each other.'

As Terry told me with a slight tremble in his voice: 'That was the scariest thing I've ever seen in my life.'" Later that evening, the dreaded blue ball returned. This time it hovered in the face of a cow. Again the globe retreated and the Sherman's three dogs, after some coaxing, took off chasing it in snarling hot pursuit.

Gwen and Terry watched as the dogs followed the glowing globe into a wooded area. They lost sight of the ball and then heard a piercing yelp. The three dogs did not return. Deciding that discretion was the better part of valor, the Shermans decided to wait until morning to investigate." The next day. Gwen and Terry found three burned circles in the woods. In the center of each circle, they discovered a greasy blob of what looked to be 'shortening or butter.'

The trees above the burned rings also had a 'scorched' appearance. According to Terry, the grass eventually grew back, but the tree limbs died. The butterized dogs were the final straw for Gwen and Terry. 'We just couldn't go on without our dogs,' Terry said. Feeling that they could no longer guarantee the safety of their children, they decided to call it quits. They would put the place up for sale and leave the hellish ranch.

"As Terry later said, 'There were some really odd things about the place we noticed when we moved in. We should have known something was wrong.'

"The ranch's previous owners, who had no children, had lived there since the 1930s. Mr. Meyer had died 15 years previously and Mrs. Meyer lived there by herself for seven more years before she died. When the Shermans moved in, they had noticed heavy dog chains bolted by each of the four exterior doors. The Shermans assumed the couple had a dog which they moved from chain to chain to keep it out of the sun. Terry inquired about the dog from a previous ranch hand. No, he was told, the Meyers had 'four huge, ferocious dogs' which they kept chained by each of the doors.

UFOS DEJA VU

"After their decision to sell out, Terry had fallen into a conversation with a group of Ute Indians who worked at the local water department. The Indians told Terry that they had formed a pool to take bets on how long the Shermans would last on the ranch. The longest guess was a year and a half. The Shermans lasted two years. A local Indian shaman friend of Terry's told him that there were tribal songs about the 'spirits and spooks' of the ranch area 'going back ten generations.' The shaman said the area was considered 'unholy ground' and was 'on the path of the skinwalkers.'"

STRANGERS COME A-CALLING

I don't know if we should consider them in the same league as the dreaded men in black, but some strangers did show up on the ranch, according to Dave, who heard directly from Terry on the matter. "Among the stream of curiosity-seekers to the ranch in the Shermans' final days was a man who identified himself as a 'Naval Intelligence Officer' from North Carolina. The 'polite' Navy man sympathized with their situation and had a great interest in reviewing their photos and videos. Another man, who wasn't' so polite, lurked around the property in a white four-wheel-drive vehicle. Terry noticed that it had different plates every time he saw it. After an angry confrontation, Terry took the man's photo. Doing a little detective work of his own, he determined that the man was an agent with the Air Force Office of Special Investigations out of Hill AFB."

As might be expected, things like this have to come to an end, and eventually the Sherman's decided to pull up stakes and move on. "The Shermans spent their last day on the ranch rounding up cattle. By late evening they were 'bone-tired.' They locked all the doors and saw their children to bed. Gwen and Terry took hot showers and then fell into a deep sleep. The next morning they awoke to find their bedding covered with blood. They both had a one eighth inch deep 'scoop mark' in the same place on their right thumbs. The ranch from hell had managed to nick them one last time.

"Selling the ranch had posed a dilemma for the Shermans. 'We didn't' want to put anyone at risk,' Terry said."

Van Eyck commented about the Shermans' dire need to unload the ranch. "Bob Bigelow's been a savior to them because he got them off the ranch. I really am impressed with the Shermans. They had chances to sell the ranch. Terry told me that a guy from Colorado wanted to buy it. Terry just didn't feel comfortable because he was afraid that this guy and his family would go in and have the same experiences. So Terry, not wanting to put any other family in that position, really had no choice but to sell to someone like Bigelow."

"At the suggestion of several different researchers, the Shermans had been put in touch with Bigelow. In September, 1996, the deal was finalized. Bigelow bought the ranch for less than the Shermans paid for it. Terry also sold Bigelow 'a select herd of cattle' and was hired on as an 'overseer' for the operation. As part of the deal, the Shermans signed a nondisclosure agreement which barred them from making any further statements about the ranch or their experiences. Meanwhile, the Sherman family had relocated to a ranch 20 miles away. After the move by the Shermans the ranch remained in the hands of Bob Bigelow and his National Institute for Discovery Science (N.I.D.S.), who turned the ranch into a paranormal laboratory. "Keep Out" signs were posted. The ranch was fenced and the gates were locked. Bigelow's workers erected an observation tower and a pair of scientists and a veterinarian were moved in. Twelve hundred letters were sent to local ranchers asking for their cooperation in reporting missing or mutilated animals.

There are some who insist that Bigelow and his team acted in a heavy-handed manner, shutting out the public as well as other researchers. On the Paracast, Dr. Frank H. Salisbury, who devoted great effort in gathering and organizing hundreds of reports with the help of Joseph "Junior" Hicks, related that he was denied access and told it wasn't a "good time" when he requested that he be allowed to visit the ranch and speak to those doing research on the property. Critics point to Bigelow's governmental and military connections and his association with those who are thought to have a "black ops" background. But all such matters are now water under the bridge, as the ranch was sold recently to an unidentified party who already seems to be more cooperative in sharing information, agreeing to the production of a major documentary by independent producer Jeremy Corbell, who was given pretty much carte blanche to the ranch as well as being able to access journalist George Knapp's voluminous files gathered while working on his groundbreaking book, "Hunt For The Skinwalker."

To move matters in an even more positive direction, the History Channel has just greenlighted a new nonfiction series which is still in its primary production stage, and whose working title is, "The Secret Of Skinwalker Ranch," that "will gain unprecedented access to the secret site of unexplainable paranormal events and UFO sightings. The Executive Producer will be Kevin Burns, the creator of "Ancient Aliens" and cable's #1 nonfiction series, "The Curse of Oak Island."

A media release clearly states that they will have "full and unprecedented access to one of the most infamous and secretive hotspots of paranormal and UFO-related activities on Earth, and will feature a team of scientists and experts who will conduct a thorough search of this infamous 512-acre property. They will attempt to find out the truth behind more than 200 years of mysteries – involving everything from UFO sightings and paranormal activities to animal mutilations and Native American legends of a shape-shifting creature known simply as, 'The Skinwalker.'"

The release from the producers points out that, "In 1996, the property was

UFOS DEJA VU

purchased by billionaire businessman and UFO enthusiast, Robert Bigelow, who used it to conduct his own experiments into the study of the ranch and its otherworldly connections. Three years ago, the property was sold to another, mysterious owner who – for reasons of his own – has, for now, chosen to remain anonymous. Except for the making of a handful of documentaries, few have ever gained official access to Skinwalker Ranch and none have ever been able to bring cameras onto the property for a television series. But that's all about to change as the History Channel uncovers the 'who, what and why' of an area that until now has remained as secretive and as forbidden as some of our nation's greatest mysteries."

In the meantime, we can continue on our merry way, joining Dave Perkins as we venture with permission past the heavy chained fence that leads us onto the Sherman Ranch, where few have ventured forth.

Shadows Of Your Mind

United Kingdom — DAVID PERKINS, EDITOR

https://issuu.com/shadowsofyourmind

FREE ISSUES - READ ONLINE

Shadows Of Your Mind is a magazine dedicated to the fields of UFOlogy and alternative research and history. The magazine aims to introduce people to the basics of fringe topics which are sniggered at by the mainstream media as controversial, kooky or just weird. We hope the magazine serves as a starting point to further personal research and enlightenment.

The Uintah Basin is a raw, rugged area of
Utah which has attracted many unexplainable events from UFOs to shapeshifters.
Blu Mountain, Utah, a famous landmark on the Unitah Basin.

UFOS DEJA VU

An eerie glow can be seen throughout the area at night as drilling rigs operate 24/7.

UFO appears to be powering up as it leaves the Wolf Creek area. www.aliendave.com/

UFOS DEJA VU

A large parcel of land, you could easily get lost without knowing your way around — as an invited guest or someone who is trespassing.
www.astonishinglegends.com

Multi millionaire Bob Bigalow gathered a team of scientists and accredited academics to study the multiple phenomena taking place on the Skinwalker Ranch.
Having sold the ranch several years ago, Bigalow is now planning to send men to the moon as part of his latest business venture, Bigalow Aerospace. >

Above: Originally called the "Gorman Ranch," in the early days Terry Sherman wanted to keep his true identity a closely guarded secret.

Animal mutilations researcher Chris O' Brien was the first in the UFO community to obtain an interview with Terry Sherman. What he discovered about the ranch then was as chilling as it remains today with its new owner who also wishes to remain anonymous.

UFOS DEJA VU

A FILM MAKER HUNTS FOR THE SKINWALKER
An Exclusive Interview With Producer Jeremy Corbell
By David Partridge

It's the height of conference season, when we catch up with filmmaker Jeremy Corbell. He's just returned from the McMenamins UFO Fest in Oregon with the legendary George Knapp to talk about his new film "Hunt For The Skinwalker." We've managed to catch him in his Northern California home in between grabbing a bite to eat and changing his socks, before he heads out to the Utah UFO Fest in Cedar City to do it all again. Riding high on the success of his last picture, "Patient Seventeen " (about Dr. Robert Lear's research into alien implants), Corbell is highly energized about this new film, a story at least 25 years in the making.

Skinwalker Ranch, for those of you who don't know, first came into the public consciousness in 2005 following the release of the book Hunt For The Skinwalker by the aforementioned George Knapp and co-written with Dr. Colm Kelleher, a bio chemist currently affiliated with Tom Delonge's To The Stars Academy of Arts & Sciences. The book which "raises the hair on the back of your neck" according to Corbell, related the unsettling experiences of the Sherman family who had bought a Ranch near the town of Fort Duschene, Utah. From the moment the family moved in they realized there was something extraordinary about the property.

Early on in their tenure they had a close encounter with an unnaturally large grey wolf on the property; lights in the sky and poltergeist activity were relatively 'mundane' occurrences in comparison and we'd encourage you to get hold of the book - if you haven't already - to get fully acquainted with the often terrifying encounters the Sherman family experienced, as well as the scientific studies conducted after Robert Bigelow had bought it off the Gormans. But more on that later.

Following the purchase of the Ranch, the previous owner 'Terry Sherman (previously identified as Tom Gorman), who had his life turned upside down for 22 months, decided to stay on at the Ranch in almost a caretaker capacity while the rest of his family moved away. "He stayed on for years with the NIDS team. The family had left the Ranch in fear, and it's hard to express to people what that is like – it's hard to understand for myself – and it followed them…" The 'It' in question is not a flame-haired clown carrying a red balloon but an unknown tall shadowy physical entity. "They moved to two different states to try and shake this thing. Maybe there's a part of them that wants to tell their story at some point… but right now, not so much. It continued to plague them, as it did the DIA agents who came to the Ranch."

"But Terry wanted to find this thing, he wanted to understand this thing. He stayed on for a long time as a caretaker to help the NIDS team hunt the phenomenon. He told them 'You need to stalk this like an intelligent wild animal' and you can. It seems to be a precognitive, sentient intelligence in some way but in another way it can make mistakes, and he knew that because he had been stalking this thing at night. One time it was pretty cold out and he'd been stalking this thing, belly crawling and laying in wait for hours, when his knee cracked a branch on the ground. There was a glowing orb, or an intelligence scanning the field. It noticed him, and it just took off in an evasive maneuver."

The property itself is located on reservation land in the Uintah Basin to be precise, and belongs to the Ute tribe, so it goes without saying that permission to film in the Uintah Basin on tribal lands requires authorization and permits from a Tribal Council. "Oh yeah, we were very lucky," says Jeremy. "My team went out there and George Knapp had created these really incredible relationships through the publication of his book, the fact that he and Dr. Kelleher had really taken it seriously. The book was very clear and didn't make light of things that shouldn't have been made light of, so there was a good relationship in place already. But, over the last few years we've had unprecedented access given to us by the Ute Nation, by the tribe itself, to not only visit the historic areas but to get access to areas such as Dark Canyon, which is where it is understood the Skinwalker lives, and then travel along Skinwalker Ridge. Historically within the culture of the tribe that's a no-go area. That's the direct path of the Skinwalker and that ridge goes right through the edge of what we call Skinwalker Ranch."

"They also let us camp at Bottle Hollow - fully permitted - film there, do interviews with people and take that footage off the Nation's land. That has never been granted before, so we are very fortunate to have this really good relationship that is still growing to this day." Jeremy's almost nonchalant mention of camping on a trail known by locals to be frequented by an actual physical boogeyman sets our spidey-sense tingling "We were there with a tribal member and tribal security. At some point during the night while we were walking this feeling just came over us – I mean I felt it, something all around us – and I looked at my security buddy and, man, he looked like he was going to crap his pants. He said: 'Yeah, I just felt something too.' Typically, though we didn't see anything, but I felt really good. I just feel good in nature. Where I live, is Pioneertown, and I can be out in the wash and there's not a single person for miles, I know I'm the most aggressive predator in that wash. Out there, on this ridge, you don't really know what's lurking. If these stories are true, if even one of them is true out of the thousands that have been told for generations, there is something there. It is intelligent and it's not necessarily benevolent." We don't doubt that for a second and pants-crapping seems like a relatively respectable alternative, all things considered. To quote the late Sonny Landham as Sioux tracker Billy Solet from the film, Predator, starring Arnold Schwarzennegger, 'There's something out there waiting for us...and it ain't no man.'

But what about the fabled Dark Canyon? The 'Lair of the Skinwalker' itself? If Bottle Hollow is a favorite haunt of the entity then surely an overnight stay in his domain would have been the litmus test of 'just feeling good in nature?' "No, we didn't camp up there, but we were allowed to film and do interviews at night in Dark Canyon. That name is fitting, man; you literally cannot see in front of your hand. I mean, even with the moon out the trees there are so dense you can't see but two feet ahead of you."

"It's a very eerie place to some people but to me it was just raw nature. Black, just total darkness. I don't mind so much if I hear things going bump in the night as I usually think there's some grounded reason for that. Although there were some owls that were going crazy near us. One of our native guides that was with us thought it was time to go when he heard that... I guess there can be a Native American association that when an owl howls a certain number of times for a number of people, when on their land I trust in the Ute culture. So we hightailed it when he said it was time to go. But, yeah, it's beautiful land, man."

For the last two years, Jeremy Corbell has had boots on the ground documenting the phenomenal story that surrounds the Ranch in the hope that something may just appear on camera. The current owner (who will remain confidential, Jeremy tells us) purchased the property from billionaire Robert Bigelow in 2016 with an understanding that the study into the strange goings on continue. "Bigelow owned it from 1996 all the way up through the Government investigation with the Defense Intelligence Agency - where they created BAASS – (Bigelow Aerospace Advanced Space Studies) - and sold it in 2016. He owned it for 20 years, studying both in a private program through NIDS and then through a coordinated study with the Defense Intelligence Agency through the Department of Defense using BAASS as the vehicle to do that study. Upon the sale in 2016, a window of opportunity occurred where this movie was able to be made and I could source footage that George Knapp had been storing.

Jeremy tells us that the film brings to light a lot of what's in Knapp's book but that it also has a few surprises – well, that's a surprise! "The new owner has gone on camera with me too – we've blocked out his face and distorted his voice of course, as he wants to maintain confidentiality and for good personal reason – and he explains, in high detail, how the endeavour, from his perspective, has been amplified and he is seeing new phases that have never been seen. So, there is interest not only by the individual who owns it, but I can confidently say that there is renewed interest by other agencies."

Huge wolves, creatures crawling out of a hole in mid air, red orbs, large pre-historic looking birds with a monumental wingspan are just some of the phenom-enon witnessed at various times on the Ranch but there is one entity which can worry even the toughest pair of military grade underwear, as Jeremy will now explain... "We were in the Command and Control Centre for NIDS, which is on the Ranch property. There was a stoic, soft spoken military man - highly trained - security individual, who's come with us for one of our guests.

"It was our last night and I'm there with a few individuals and it's like 3.30 in the morning. I was coming in from outside and this military individual is sitting at a table. He's the only one facing me, and as I come out through this dark area from the back into the kitchen, I see his eyes, warning me as if to say 'sit back at the table,' but he doesn't say anything. After a moment or two he just says: 'OK guys, let's go'. We get in the car and drive out past the gates of the Ranch. At this point it's about 3.50 in the morning and he stops the car, turns it off, turns to me and he says "I've never seen anything like that in my life." And I'm like "What are you talking about?" This is a guy who is not prone to exaggeration, he's a highly trained military observer. And he says "When you walked from that back room there, I thought it was Matt (Adams, one of the cameramen). When you walked into the light into the kitchen area, behind you, clear as day, I saw a tall dark figure, a six foot something tall shadow person that looked like a physical tangible smoky thing. I thought it was my eyes." and he then says "And then 'poof' it just evaporated behind you but it was following you. Right behind you." I didn't see it, I didn't feel it, but what he didn't know is that this apparition-type being – and this is some-thing that's never been made public – has expressed itself in that exact location before. But he was not told about that, I had never told him and George had never told him. So I found that highly credible, however I did not personally see it: I think I just got a hitchhiker." [laughs].

OK, so the shadow being did a Keyser Soze but the question is, did Jeremy bring it back with him? "I don't know, man! People have come over to the house and they've told me they've seen something lurking but... I don't know. Every-body else seems to experience the paranormal: I don't experience anything! I wish I could report to you that I had a completely powerful, paranormal UFO ex-perience, but it's just not the truth. Personally I did not have a close encounter with anything unknown. I did see lights at the second homestead, really bright, lighting it up from a distance and there's no artificial light out there, but I don't know what that was: it's just 'something'. No, trust me; I would have held up my cell phone and captured it for everybody. But that's the big joke: people are afraid of ghosts and stuff but all they gotta do to never be haunted by something, is to carry a cell phone everywhere they go because no ghost ever likes being on camera. It's like kryptonite to the paranormal!"

"But remember how the scientists described the phenomenon is that it messes with you. It shows people specifically certain things and excludes other people from that display, it's selective. This is what it does. When the DIA individuals came to the Ranch to assess it from the very beginning, why they took it seriously was because they had displays, to them, directly." As we ponder this exhibition to a highly trained military individual a thought pops into our head, and we pose the question of whether this particular entity targets security or military personnel as it sees them as the strongest members? "I think it's like it's a provocateur. 'Oh you want to come study this? OK, well I'm going to give you a display Mr. DIA!' The way that it seems to target, the security that are there, maybe it's matching their intent. The phenomenon seems to create performances when a few things are done in the area and on the Ranch. Digging up the earth, the arrival of a stranger, making noise, that sort of thing. It seems to target security personnel often and it's a very selective intelligence and plays a game of cat and mouse. It's always toying with you.

"Psychologically this thing will target people, target their families, follow them home. I can only assume it's because it's a provocateur, the intelligence that resides or intelligences that reside at the Ranch - that includes UFOs but it is not limited to that." We agree but given the history, whatever is at the Ranch is up there with any monster the Hollywood machine could dream up; Freddie Krueger, Jigsaw, Harvey Weinstein...

When it came to actually documenting and doing the actual hunting - and being used as paranormal 'bait' by George Knapp - Jeremy makes it clear that each time they have gone to the ranch, it was to experience and engage the phenomenon rather than try to capture it on camera. On one occasion the previously mentioned friend pretty much summed up their presence at the ranch "Me and Robbie Williams..." Wait.. Robbie Williams...? As in Rock DJ Robbie Williams? Heart-throb to millions of a certain age, Robbie Williams? "Yeah, Robbie's really, really curious about the paranormal and UFOs. We were out walking at about 3 AM and he turns to me and says 'So the NIDs team was here on the Ranch attempting to document the high-strangeness for seven years? And they only documented about a hundred dramatic encounters throughout that time?' He looks at me and we're in the middle of the ranch in the pitch black and he says 'So we can't expect to see anything? I mean we are truly hunting for the proverbial paranormal needle in a haystack?,'" and when he made that comment I started laughing because he's right, you need to invest the time.

UFOS DEJA VU

"You can't just turn up at the gate at Disneyland for an hour and expect to experience Magic Mountain! It takes dedication. Scientists and intelligence agencies have studied the Ranch for a multitude of years and they only catalogued a certain number of events." Robbie's a good person y'know, sober minded and direct. I don't think he's worried about coming out again in this film saying he's interested in this stuff. There will be a 20 minute bonus material piece of him, accompanying the launch of my film, where he's talking about his UFO and paranormal experiences and featuring one of his songs, Arizona. He's in the film because he has a sincere and unwavering interest to understand the truth, and a creative drive to explore the unknown. He's one hell of a guy, I admire his genuine curiosity."

Talking about the studies made by NIDS and subsequently BAASS, what kinds of tests did they undertake? We're guessing they tried communicating, attempting a dialogue with the phenomenon? "Yeah, and they did some pretty incredible stuff. In-depth scientific and technical reports aside, some of their experiments were fascinatingly playful. One of which was using 'jacks', the toy when you're a kid and you have jacks and a ball? Well they put some Jacks in an order on a shelf in a controlled and monitored environment. When they checked sometime later, those jacks had been moved but they couldn't determine if there was a message in it or they had just been moved. They tried to establish a line of communication with the phenomenon, through puzzles and games and that sort of thing."

As we try and get our heads around the fact that Robbie Freaking Williams was at Skinwalker Ranch – although a tweet he made in December did kind of give a big hint – and that NIDS were trying to play children's street games with unknown entities, talk comes full circle to the man who began it all: George Knapp. After sitting on an incredible amount of data for a quarter of a century he must feel a real sense of accomplishment in seeing this documentary being made public? "Yeah, he's stoic but I can only guess he's excited to see this finally be told in such a large medium. He's the reason why I'm able to tell this story because I pestered him and pestered him and he let me in to see the archive, documents, photos and videos. I was like 'Holy shit! The Akashic Records of Skinwalker Ranch!' He agreed, and we got authorization because, with the change of ownership of the ranch it was a new era but there is also the matter of one's word. So this never before seen footage was allowed to be implemented into the film."

It must have been a challenge to get the pace and flow of the story into the version people will be viewing? "In this film, I have started from the beginning to tell the experiences of the Ranch. And I dedicated the beginning and the end of the movie to clarify what's actually going on, with the agencies and funding and how it all relates to the recent UFO revelations by the Pentagon. I want people, the general public, to start to understand what Skinwalker Ranch is. It's a game changer"

UFOS DEJA VU

With a running time of around two hours seven minutes, cut down from "over three and a half hours" there's a definite buzz surrounding the film. "I can't wait to see it on the big-screen, man! I've looked at every frame of the film, but it's totally different when it's out. There's so much going on at Skinwalker... it's not just UFOs. It centers around UFOs and the Government study, but we're talking portholes with creatures coming out, UFOs, orbs, animal mutilation, poltergeist activity. This is close encounters of the third kind with every incident; it's people engaging entities. And there's more to come that will set the record straight about what actually happened at the Ranch. For the first time in history people will be able to see footage from the time of the active investigation, thanks to George Knapp's archive. This is the documentary you were never supposed to see."

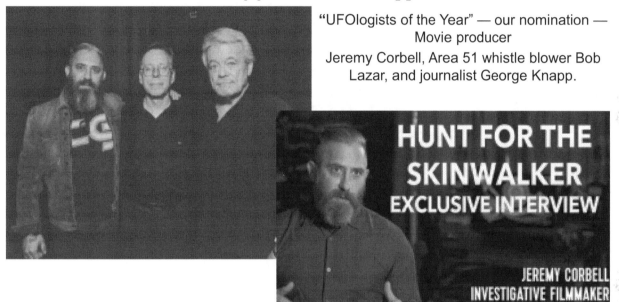

"UFOlogists of the Year" — our nomination — Movie producer Jeremy Corbell, Area 51 whistle blower Bob Lazar, and journalist George Knapp.

Jeremy Corbell Courtesy Openminds.tv

The reviews and interviews for Jeremy Corbell's production have received notable attention. Here he is seen on a widely viewed news broadcast.

UFOS DEJA VU

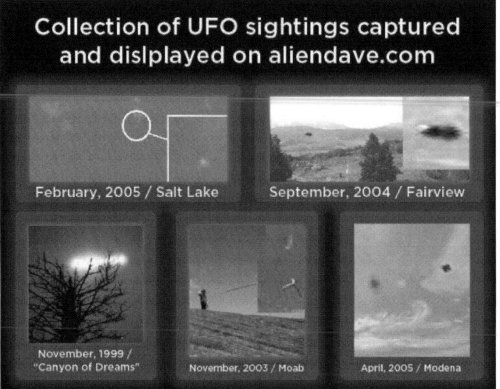

Collection of UFO sightings captured and dislplayed on aliendave.com

February, 2005 / Salt Lake

September, 2004 / Fairview

November, 1999 / "Canyon of Dreams"

November, 2003 / Moab

April, 2005 / Modena

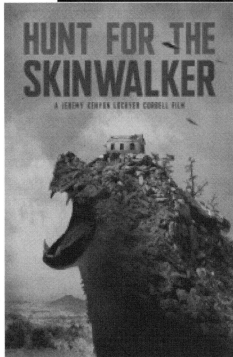

< Based on the best-selling book by George Knapp & Dr. Colm Kelleher, filmmaker Jeremy Corbell is launching his new film, HUNT FOR THE SKINWALKER. With distribution by The Orchard, this documentary is an intimate and unnerving portrait of the events surrounding the most extensive scientific study of a paranormal hotspot in human history.Order from HuntTheSkinwalker.com or Amazon.

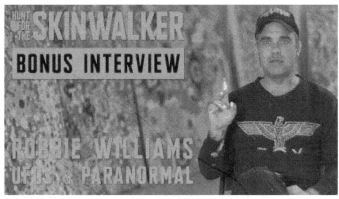

Right: British pop singer singer Robbi Williams is looking for aliens in new documentary, "Hunt For The Skinwalker."
Exclusive Interview — https://www.extraordinarybeliefs.com/news-3/robbie-williams

Trailer — www.youtube.com/watch?v=Gi6XNjiEHCc

UFOS DEJA VU

AN EXCLUSIVE INTERVIEW WITH "MR. SKINWALKER,"
GEORGE KNAPP
By David Partridge

PUBLISHER'S NOTE: I really appreciate doing Coast to Coast AM. All the hosts are fabulous and the program staff – thank you Lex and Greg – always does its utmost to post as much information as possible about their guests, including links to websites and Amazon pages. You can even add photos of what you are planning to talk about that evening and everything is kept permanently on the site so that new listeners can be bought up to date on a guest's ongoing activities over the years. Believe you me, that's a plus when you are trying to promote a new book and want to reach the largest possible audience. And Coast does have a large audience – an estimated 3 million listeners some nights. Above all else, I enjoy chatting with weekend host George Knapp as he seems so keyed into the subjects we write and publish about.

As background – born April 18, 1952— Knapp is an American television investigative journalist, news anchor, and talk radio host, who has been recognized with Edward R. Murrow Awards, Peabody Awards, and numerous regional Emmy Awards. A longtime fixture in Las Vegas media, he works at KLAS-TV.

Another thing I like about GK is that he seems particularly passionate about the UFO topic, which is rather unusual for a professional journalist (though the list of reporters now "intrigued" about the subject may be growing). As an example of George's early philosophy on UFOs we pulled part of an interview "For The People" host Chuck Harder conducted a number of years back when Knapp was just getting started investigating the topic and had just hooked up with the controversial Bob Lazar; this would have been pre Skinwalker days.

"Our research actually started about two and a half years ago, a fellow named John Lear, the son of the guy who invented the Lear Jet, brought some of this information to our attention. In May of this year, Mr. Lear introduced us to a fellow who claims to have worked at a secret base designated S-4...on a top-secret Nevada test site...the fellow said that he worked on flying saucers, that the technology was not from Earth, and we interviewed him live in silhouette (of course, that was Bob

UFOS DEJA VU

Lazar, who Jeremy Corbell has done another film about). The response was incredible. We got response from Japan, and parts of that interview aired on the radio in six different European countries. So we decided with this much interest we might want to take a look at the subject in more depth.

"We started doing that and the first thing we found out is that really UFOs have not been given a fair shake by science, by government, by religion and especially by journalism. Millions of people have seen UFOs, millions more believe. I think the latest Gallup poll shows about 70 percent of college-educated Americans believe that there's something to it, but because of the tabloid aspects... 'The Girl Who Gives Birth to 52 UFO Babies' — kind of things – people have shied away from it...Serious people have shied away. Scientists, although they might be interested in searching the universe for radio signals really don't want to look in their own back yard — they can't get grants...people would laugh at them.

"Journalism—the coverage is generally condescending and quirky, especially by the networks, as in the coverage of this UFO that supposedly landed in the USSR, people making fun of it...so, we figured out that millions of people want to know as Roy Neary, the guy in the 'Close Encounters' movie said, 'What's going on?' So we started investigating it.

"The focal point of the story being this fellow who said he worked at S-4. We broke his identity last Friday. His name is Bob Lazar, he is a former scientist who worked at Los Alamos National Labs, he is a physicist...we did a lot of checking on him and found interestingly enough that his life was disappearing around him. In other words we called Los Alamos Labs and they said they never heard of him. We called MIT where he says he went to school and they had never heard of him.

"We called for his birth records and they had disappeared...as if someone was trying to make him a nonperson. We did, however, confirm some of the information that he had given us...we found newspaper articles from Los Alamos indicating that he had indeed worked there...we found an old telephone book from the lab with his name in it, which gave him a certain amount of credibility in our eyes. The story he tells is an incredible one. He was hired to work at this area called S-4 on the test site, he was flown up to a place called Groom Lake — taken by bus with no windows to S-4...the base is built almost to look like its part of the desert with sand covered hanger doors, he goes inside and he starts reading these briefing papers dealing with UFOs! Pictures of UFOs on the walls, pictures of aliens, autopsy reports on alien bodies...things of this nature — he's pretty amazed. Then he sees the discs. He says there are nine of the discs up there, they are powered by an antimatter reactor which produces its own gravitational field...technology that does not exist on this planet, and the interesting thing...he thought for a while that perhaps it was just an advanced secret scientific project that our government is pursuing until he looked inside one of the discs and noticed the small furniture...all the chairs were built like for children, and then things started coming together for him. Are you with me?"

But it's about time we move on with the topic updating to just a couple of month's ago GK's thinking on UFOs with the interview conducted by Denis Partridge.

* * * * *

We couldn't end an article about Skinwalker Ranch without speaking to the man who was jointly responsible for bringing it to the general public's attention. Taking a break from unraveling the various threads of the Pentagon UFO debate, among other things, with his 'I-Team' from LasVegasNow, investigative reporter George Knapp kindly took time out to answer a few of our questions.

QUESTION: Has the atmosphere around the Ranch changed and evolved since your first visits there?

GEORGE KNAPP: The most obvious change involves security. In 1996, around the time when the Ranch changed ownership, there were a couple of newspaper stories published which mentioned UFO sightings and animal mutilations but those reports included very little about the most bizarre, truly disturbing events that had occurred during the 20 months when the "Shermans" lived there. And of course, there was no mention of the even stranger events that had not yet occurred. It wasn't until my articles in 2003 that the public learned about the much broader scope of inexplicable activity, and that is when problems erupted with trespassers and vandals and assorted nincompoops. After the book 'Hunt For The Skinwalker' was published in 2005, things really hit the fan.

Intruders not only increased in number, but also grew bolder and more reckless in the kinds of things they would do on the property. Ever since then, it's been necessary to maintain a strong and visible security presence on the ranch, and I think that has caused a backlash of sorts. It has affected the level and types of unexplained activity there. The one and only thing we have learned for sure about the Ranch is that it is interactive. It seems to respond to outside stimuli in unpredictable ways. It revealed itself to the NIDS team for the first few years, showed them some amazing and befuddling things, then went underground. Fewer and fewer incidents were reported, and very few were the kind of thing that could be investigated to any degree.

The ranch was essentially dormant for a few years, though there were ongoing eruptions of weird stuff elsewhere in the basin and on neighboring properties including Ute lands. It was as if the entity or intelligence that used the ranch as its playground or way station suddenly grew tired of being pursued, or was uncomfortable about having the tables turned by the NIDS team. It did some very dramatic things that seemed to say, enough is enough, this game isn't fun anymore. And then it went into hibernation, or it went underground, or it went somewhere else to mess with other people who were more easily frightened than the NIDS guys. As soon as the property was sold in 2016, things started back up again, according to the new owner and his team.

QUESTION: How has your understanding of the phenomenon evolved since writing your book with Dr. Kelleher?

GEORGE KNAPP: Because of my friendship with Robert Bigelow and later with a few other principal folks at NIDS, I was allowed to know things about the Ranch investigation pretty much from the beginning in 1996. So it became clear right away that this went way beyond mere UFOs, but it didn't fully sink in until I was allowed to visit the place for the first time. I had covered UFO stories for more than a decade by then and although I had started out like many others by embracing the then-dominant paradigm that flying saucers are nuts and bolts craft piloted by ETs, it wasn't long before that explanation no longer made sense to me. It simply did not fit with all of the stranger aspects that always seemed to be hovering on the periphery of UFO sighting reports.

After reading all of Jacques Vallee's stuff, and pretty much everything else I could get my hands on, I was looking for some kind of a model or theory that could accommodate the truly weird stuff. At the time, I thought it was my own original idea that maybe those bizarre events on the periphery of the UFO experience - the things that made people way more uncomfortable than encounters with mere space aliens - were actually the heart of the matter, and that the saucers and other UFOs were sort of a distraction, part of a learning curve. And then, holy crap, I got the chance to learn from and interact with the deepest thinkers to ever tackle this stuff. Bob had put together the best minds in the world, the people who had thought more about these matters than anyone anywhere, including his right-hand man Colm Kelleher, Jacques Freakin' Vallee himself, geniuses Hal Puthoff and Kit Green, the courageous explorer Edgar Mitchell, John Alexander, John Schuessler, Al Harrison, and several others who didn't want their names made public. These people had devoted much of their lives to figuring out esoteric matters that most of their colleagues considered to be bullshit. Along the way, they risked everything - their jobs, their reputations, their security clearances, everything.

It was an astonishing time for me. I was allowed to be a fly on the wall when these great minds got together to discuss, debate, strategize. They let me listen and learn and sometimes contribute, allowed me to overhear things that were extremely sensitive and way over my head, with the understanding that I would not report or reveal any of it, until or unless they gave the green light. It was an arrangement I was more than willing to make, (and would do so again) and because I kept my word back then, friendships were formed and trust was established. Although I could not and would not spill any of the info I learned in confidence, these interactions helped me process other information that came my way.

UFOS DEJA VU

It gave me a unique vantage point in the larger UFO context. It helped me in the evaluation of assorted loud voices, boisterous claims, and grandiose cases that popped up in the wacky world of Ufology from time to time. The great minds on the NIDS SAB (Science Advisory Board) had many debates about the Utah Ranch. It became clear that whatever intelligence was operating there, it acted contrary to the classic assumptions about supposed ETs. This was something else.

Eventually, the NIDS investigation of the Ranch changed my view of pretty much everything, right down to the nature of reality itself. This is a genie that can't be stuffed back into the bottle. In the years since the book came out, we've collected so much additional information about related phenomena, and it seems obvious that while the Utah Ranch is special in that the unexplained activity occurs in such dramatic concentrations at times, the Ranch is not entirely unique. It is special, in part, because of the amount of scrutiny it has received. The same things are happening elsewhere, maybe everywhere, to one degree or another. Eventually, I came to understand a very uncomfortable idea: the idea that another intelligence lives here. It is not an intruder from somewhere else, not a traveler just passing through. I think people will have trouble getting their heads around the idea that we humans are living in someone else's world, not the other way around.

QUESTION: Will the results of research conducted by both NIDS and BASS be made available for public scrutiny?

GEORGE KNAPP: The NIDS material is pretty much out there in the public. The Skinwalker book is the story of the NIDS study, and we included all of the major events that occurred between 1996 and 2003-ish. There are quite a few smaller, less dramatic incidents that were not in the book, and many things that have happened since it was published in 2005, but in general, we didn't hold back any big stuff. Colm Kelleher was the lead investigator. He is familiar with everything that happened during the NIDS study. He is not only my coauthor on the book but is a good and trusted friend. I know there are many people who assume there is a huge vault filled with secret files and tapes, but as far as I know, that is not the case. NIDS did record an enormous amount of video. Cameras and recorders operated 24/7 for years, and there are some images of anomalous objects, but for the most part, they are not compelling.

There are orbs, lights, floating in a sea of inky darkness, just illuminated blobs with no points of reference, and there is no way to draw any conclusions from that footage. There were some ghostly images that could be seen on the video monitors—things that looked like tall buildings, a few daylight shots of fan-shaped lights near the ridge, but what can be said about them? As for evidence, anyone who has read the book and does not accept the central premise that this intelligence played games, seemed to anticipate what the investigators were going to do, avoided detection, messed with their minds…well, if that is too much to swallow, so be it. That's the way it was. There were reports from the field, sent by personnel on the Ranch, and later some emails about particular events, and while I would love to see them if they still exist, they are not going to solve the mystery. There were several distinctly physical events that occurred, and I have heard people say they demand to see the evidence, but evidence is relative. For instance, compasses would often spin out of control. Batteries died. Entities were seen, sometimes only in the infrared range. Is that physical evidence? The entire corral was magnetized during the incident when the four bulls were stuffed into a trailer, but what kind of report - or evidence - would convince skeptics that these events occurred? Most of the key events from the NIDS era are told in the book and now have been brought to life in Jeremy Corbell's documentary.

As for the BAASS study, that is another matter altogether. Back in 2008, just one week after Bob Bigelow signed a contract with his new government partner, he agreed to be interviewed by me on Coast to Coast radio. In that extraordinary program, he pretty much laid it out, left a trail of breadcrumbs for anyone who was paying close attention. He announced to the world that he was launching BAASS, that it was created to investigate unusual phenomena, and that he had "a partner." He never named the partner, but it was a pretty big clue that was universally ignored. I could not say anything about it at the time, and was not sure it would ever be made public, but some of the beans have been spilled, and I tried to explain some of it in the movie.

QUESTION: With the release of the film, I guess there's a chance there may be a potential unwanted increase in paranormal 'tourism' to the area?

GEORGE KNAPP: It might be the only prediction we can make about the Ranch. Yes, people will come, but they're already there, every day and every night. The fact that it is a long way from any major population centers is a mitigating factor. People have to really want it to travel that far. The new owner has spent a lot of money on security measures, including technology that is not beyond the norm. We did not want to cause any additional headaches for the owner, so we sought his opinion before proceeding with the film. He thought about it for awhile, then gave the green light. Jeremy somehow talked him into appearing in the film, though his identity is not revealed. The Ranch is ready for intruders.

QUESTION: How do the Ute oral traditions of Skinwalker activity in the area differ from what has been seen in the last twenty years?

GEORGE KNAPP: I remember the day when I first heard that word. It was on my first visit to the Ranch. We were chatting with UFO expert Junior Hicks about the history of all of the unusual phenomena in the basin, and he pointed to the sandstone ridge that runs along the northern boundary of the Ranch. He casually mentioned that the Utes referred to it as Skinwalker Ridge because, in their oral traditions, it was a place that was "in the path of the skinwalker." I got a little chill when he said it, though I had no idea what it meant. When he explained the context, I thought to myself that it would make a great title for a documentary. Back then, I was hoping to produce a film about the Ranch, not a book.

A few years later, after the film project had been shelved, Colm and I started working on a book. We had a lot of trouble finding information about skinwalkers. There was very little online, and when we reached out to tribal governments, including the Utes, Navajo, and Hopi, we were shut out. No one wanted to speak to us about that subject, and most of the people we called would not even say the word. We managed to piece things together, in part by including testimony from witnesses who had cooperated with NIDS, and also by interviewing people who had lived with these tribes but were not members. The general story about skinwalkers was creepy, but we knew from the reactions we had received that tribal members took this very seriously, and that is how we approached it as well. In a nutshell, skinwalkers are evil shapeshifters, sorcerers, killers. Some Utes believed the Navajo had unleashed a skinwalker curse on the tribe

This story about a skinwalker/shapeshifter eventually morphed into an umbrella explanation for many of the strange events the tribal members had seen in and around the Ranch since the 1860s. Junior Hicks told us tribal members were admonished to stay away from the Ranch. Now, 13 or so years after the book came out, the term skinwalker is all over the place. Back then, it was mysterious, spooky, but it also worked as an umbrella term for what NIDS was trying to do. They were never literally hunting for skinwalkers but they were hunting for a broader explanation of weird stuff that was happening in that area. We used the word in a symbolic context, another way of saying it was a search for the unknown or unknowable.

In the years since the book came out, we have been successful in getting to know some members of the Ute tribe. They have explained to us that the skinwalker story is one of many explanations for what is unfolding at the Ranch. Another possibility is that the Ranch property is a place where ancestor spirits enter this world. They are open to the idea that aliens or interdimensional beings use the area as a gateway of sorts, based on personal experiences many of them have had. I harbor a suspicion that our Ute friends have other ideas about what might be going on, but are not yet ready to share that information.

* * * * * * * *

Hunt For The Skinwalker by Jeremy Corbell is available to watch on most streaming platforms. Jeremy's latest film **Lazar: Area 51 & Flying Saucers** is also available to stream now.

This interview first appeared in issue 4 of Shadows Of Your Mind magazine and is reproduced with the kind permission of the publisher. For back issues of the magazine, visit issuu.com/ShadowsOfYourMind.

UFOS DEJA VU

Preparing for a broadcast, George Knapp is a frequent weekend host on Coast to Coast AM.

Expecting anything might happen, George Knapp is on the grounds of Skinwalker Ranch.

UFOS DEJA VU

Back in the day: George Knapp during his first visit to Roswell, New Mexico.

George Knapp holds his Peabody Award aloft to the crowd.

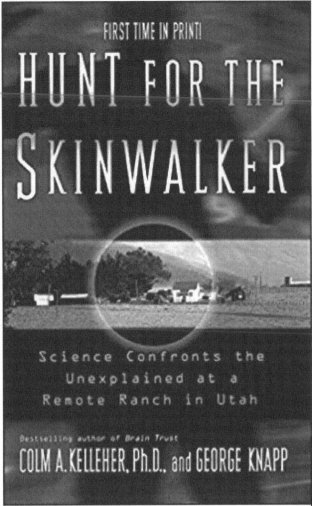

"The Hunt For The Skinwalker" provided shocking information about what was taking place on the Sherman Ranch.

Researcher/editor Dave Patridge was determined to get to the bottom of what was happening at the ranch when he published this edition of his on line magazine.

UFOS DEJA VU

THE LADY AND THE SKINWALKER
By Timothy Green Beckley

PUBLISHER'S NOTE: With her proximity in the state of Utah, Erica has had a persistent interest in the Skinwalker Ranch, a location of regularly reported paranormal activity, where she has witnessed unexplainable phenomena herself. Erica has assembled several decades worth of case reports of mass UFO sightings, alien abduction reports and animal mutilations from her years of field research in and around the ranch. She has studied at length the papers of the late phenomenologist Dr. Frank Salisbury of the Department of Plant Science at Utah State University.

Salisbury contributed important work on UFO reports in the state with the help of "Junior" Hicks, a well-respected UFO investigator. Erica Lukes is an artist and host of her own radio show, "UFO DECLASSIFIED," heard on KCOR Digital Radio Network. Since childhood she has been fascinated with imagery of how vehicles and beings from space might appear and she pursued this interest with determination.

This, coupled with her long-term interest in strange phenomena of nature, led her to take her first steps into radio in 2014 by developing her own programming and interview techniques of known figures in this topic. UFO reports became her specialty and she joined the Mutual UFO Network (MUFON) in 2014 as a field investigator. She eventually became its State Director for Utah. Within MUFON, she also became an associate producer for the MUFON Communications Team and a part of the MUFON Experiencer Research Team, having been cited for recognition of this work at both the 2014 and 2015 MUFON Symposia.

Additionally, Erica was a team leader for "Project Orange," a program to study so-called "Balls of Light" phenomenon that have ramifications in real world science in the form of ball lightning and other peculiar luminous phenomena. An offshoot of this was research into the luminous ball phenomena in Hessdalen, Norway where she conducted on-site research with the guidance of Hessdalen scientist Dr. Erling Strand. Hessdalen has been the site of strange luminous light ball sightings since the 1980s. The following is an interview we did on our "Exploring The Bizarre" podcast as relates primarily to the Skinwalker Ranch and Erica's in-

UFOS DEJA VU

volvement with what is going on out on the Basin.

http://www.ufoclassified.com/erica-lukes

* * * * * * *

BECKLEY: We have been seeing a lot of you on TV recently, for example on the Travel Channel.

LUKES: You know what? It's great. It's always nice to be involved with things I do on the Science Channel, "NASA's Unexplained Files." I'm in the new season and doing a few other things. And it gives me a platform to reach a broader audience, which I love, because I get calls and messages from people all over the world who are sharing experiences with me. Also, I get researchers who are sending me their archives, donating archives to me so that I can preserve history.

BECKLEY: Well, how did you get involved with shape-shifting and Skinwalker Ranch?

LUKES: Utah is my "hood." I've lived here my whole life. And I have had really profound experiences here. Since I was a child, I've had experiences. But in 2013, I had a series of sightings, and so I began to pursue research.

BECKLEY: What did you see exactly?

LUKES: I saw over the Oquirrh mountain range and I had a beautiful view of the Salt Lake Valley, I saw the amber orange spheres, and they would hover over the ridgeline for over 45 minutes, And then one would appear to drop out of the other, move around in a circle and then they would move off. And so, being 80 miles away from Dugway Proving Ground and being a community activist and a concerned citizen, I'm like, what is this? I want to know.

And so I began learning where the flight corridors for Salt Lake International Airport were, and where our restricted airspace is. We have some of the most restricted airspace in the world with the NSA being right here in Dugway Proving Ground and the Utah Test and Training Range. And so I learned about all of that and I began to—like hopefully most credible people in this subject—you learn to rule out mundane objects, which I did. And at the end of the day, what I witnessed was remarkable. So I joined MUFON and then really began to dig into specific areas in Utah: Skinwalker Ranch and the Basin, which is one of the most remarkable places on the planet. But there are other places just like it. In Yakama, in Piedmont, Missouri.

Dr. Harley Rutledge wrote a great book on that and did a study back in the 70s. So I went out and began to interview different families in the basin who had remarkable experiences and were in close proximity to the Ranch. And I've been cataloging different reports. I spoke with Dr. Frank Salisbury. I spent a week up in the University of Utah archives going through Dr. Frank Salisbury's work, which most people have not seen. It was amazing. And then I also spoke with and interviewed Junior Hicks. I've been doing a lot of work on that area because it is absolutely remarkable. And not only are there really profound things taking place, but then you've also got this aspect of people that have gone in to try to make money on the location who have manipulated public perception and created some very negative and destructive things that go on in the community that are putting people at the ranch and in the area at risk. So the whole thing's a freaking trip.

BECKLEY: Well, how far back do these incidents go? And is there a similarity in all of them? You described this light phenomenon that you saw, but are any of them actual craft, do you think? Are there similarities or dissimilarities?

LUKES: You know, there are structured craft that have been photographed and witnessed. There are areas in the Basin where they're seeing shape-shifting creatures. I interviewed Chris Marks, who worked for Bigelow Aerospace, on my radio show and he was at the ranch for six years. He witnessed shapeshifters, skinwalkers. He witnessed different creatures. Wolves that were shot at almost point blank range and they wouldn't drop. These are the types of reports, also cattle mutilations, I mean, these things have been taking place for a long, long time. We're talking decades if not longer. And when you reach out and do work and you talk to the people in the community, you get reports of the same types of phenomena taking place for a long time.

But, more than that, it's not just at Skinwalker Ranch, Tim, as you know, because you've been in this for a long time and have more knowledge than a lot of people. I respect your work. I just want to throw that in there. As you know, when you go to these places, different places in the world, they're all seeing the same types of things and recording the same types of things. We have to do our part to make sure we're getting out there and cataloging these reports.

BECKLEY: Have you, or has anybody, ever had the opportunity to speak to the − I was going to say the original owners of the ranch, the Shermans, but I'm sure they were not the first people to own that property. Have you ever spoken to them?

LUKES: I have not had the pleasure of speaking to them. I would love that. I know that they're very guarded now and they really want nothing to do with it. Because I think that was a very traumatic experience to have, all of a sudden. They're scared out of their minds when they're witnessing these things and then you've got, all of a sudden, the media descending on them. Zack Van Eyck, who is an incredible reporter, wrote groundbreaking articles in The Deseret News about the ranch and about the Shermans. They endured a lot of ridicule in the community, and after they left I think they decided that they probably didn't want to have a lot to do with it.

BECKLEY: Have you ever been on the ranch yourself?

LUKES: I have. I have.

BECKLEY: And did anything materialize for you?

UFOS DEJA VU

LUKES: You know, it was a great experience. Hopefully I'll talk about that. I've actually signed with the current owners an NDA and I am working with them and with a lot of things. I respect the work that they're doing. They're doing incredible work. And I really believe they're going to take what Bigelow did and move that forward a hundred times. But I will talk about the experience that I had when I went up there to interview Junior Hicks and I was right on the border of the ranch. It was dusk. There was a lightning storm rolling in. I was there with a friend. We had a great vantage point of the ranch, so we set up our equipment. By the old homestead I noticed a light that appeared to be shining up on the mesa. I watched it, and I thought, okay, there's a road here. I've got to rule out A, B, C and D. What is this? Then we watched as this light traveled across the underside of the mesa and then it appeared to both of us that this light moved out across the grass where the bait pens are. And it appeared to expand and it looked like an illuminated mist.

We were sitting there, trying to put things through their paces. This mist moved towards us and then around us. The temperature dropped about ten degrees. The sound became very muffled. Then it would retreat. Also, during this whole time period, I noticed that above where Skinwalker ridge is, where according to some of the best researchers in the world, these craft appear to move in and disappear and then come out. But I noticed an amber-orange ball of light and I photographed that. And at the same time there was this amazing lightning storm blowing through. So it was just absolutely one of the most stunning and surreal experiences I've ever had.

We had some experiences before we went there but within an hour's span and after we left. Still, sometimes, when I think about that experience, it was so mind-blowing to me to witness what we did. I sometimes have a hard time putting it in words, but I'm very grateful that I could take photos of it. I have video of the exchange, our dialogue, and all of these things.

BECKLEY: Dialogue? With whom?

LUKES: I was with a friend of mine, and so we were out there together. And had talked to another person. I've got a person who lives out there that I've interviewed his mother and grandmother for six years. They've had experiences. And he gave me a specific location to go to so I would have the best view of the property. So I was there. And, literally, within ten minutes of us getting out of the car, this proceeded to happen.

BECKLEY: And what precisely happened?

LUKES: It was the ball of light that was illuminating the mesa, the mist that formed.

BECKLEY: Any sightings of shapeshifters or creatures? Or spacemen?

LUKES: Praise the Lord, no. I think that what we witnessed was enough. And I will say – and I haven't really talked publicly too much about my experience – but I will say that I felt, as did the person I was with, that whatever this was, was almost like a security system for the ranch. Or for this area. So it was coming out and sensing us, and that sounds nuts, but those are my intuitions – that was what I was feeling – and then it retreated. And this happened several times. So we did at one point hear something or somebody walking behind us. You could hear that in the gravel where we were.

BECKLEY: What are they trying to "secure" exactly? What makes this property so valuable to any entity, whether it's a normal human being or some visitor?

LUKES: There are sacred places all over the world. There are places where – and I believe indigenous people have understood this – they've utilized specific locations that seem to have more energy there. They manipulate this, they protect it, and I think this is one of those places. There have been significant injuries on the ranch, and I believe that those injuries happen when something is disturbed in the area. It's much like a haunting. If you're looking at people who have significant hauntings in their home, and they're doing renovations or construction, activity appears to increase. So I don't think this place is necessarily unique in that regard. Like I said, I believe this is happening all over the world. In specific locations, like Norway, like Missouri, like Yakama. And so what do we do as researchers to find out what the commonalities are? And how do we get funding to do good field research and then make the connections? And instead of sitting behind our computer and investigating a single case that we think is going to be the end-all-be-all and prove that we have other lifeforms moving around us. That's where I feel we need to move research forward.

BECKLEY: Now, after having been involved with this for quite a number of years, what are your conclusions? Do you have any conclusions to share with us?

LUKES: You know what? I don't. I wish I did, because it seems like it changes. The more I learn, sometimes, the more questions I have. I will say that I feel that this isn't necessarily extraterrestrial. I think there are aspects of all sorts of things that are interacting with us, and some people are more tuned in and can have experiences. And what it is, I have no clue. What the origins are, I don't know. I don't think anybody does. And I think it is, to me, pretty suspect that people are reaching for the extraterrestrial thing when in actuality we don't know.

BECKLEY: Do you think it's all tied in? Or are we talking about a place that just has so much energy – for lack of a better term – that a lot of different phenomena are taking place?

LUKES: Yes, I think that's exactly it. In my interview with Chris Marks, who I mentioned was an investigator for Bigelow, and he was there probably longer than any other investigator, six years. He mentioned in the interview that by the old homestead there appeared to be a certain type of phenomena. And then you got over by the river, where it was completely different. And then you got up on the ridge and again there was something different. And so it appeared that there were different energies in this environment and they were commingling or not. But there were different types of things, so you could say that at the old homestead we've got ghosts. Or spirits of departed loved ones that used to inhabit the homestead. And then you have shapeshifters, perhaps something like that going on. And then perhaps at another location it could be extraterrestrial. But it is a place where you've got all sorts of things happening.

BECKLEY: Now, how about under the auspices of the new owners? Do you know if there's anything recent that's taken place there?

LUKES: Yes, there is absolutely activity, I know, but some people claim the activity has died down and that is absolutely untrue. There is a lot of activity currently taking place. I speak with the owner frequently. The History Channel just greenlighted a new show about some of the things that are going on there, which is in my opinion great. Because they will be able to get funding to continue some of the work that they're doing. Right now they have more up there and more going on with regard to research than most places that I know. Maybe Norway could be an exception, but they're really doing good work. The nice thing about the new owner is that he is not beholden to the government of the United States, which I think is huge. So he doesn't have to dance around or do this and that. He can really do research and then eventually share his findings with the public without being beholden to the government.

* * * * *

PUBLISHER'S NOTE: Erica is among the new breed of podcasters. Her show—UFO DECLASSIFIED—is top notch especially if you are into the topics covered in this volume. It is broadcast live on Friday evenings over KCORradio.com where you can find all her shows archived – as well as ours, EXPLORING THE BIZARRE. Erica has a number of shows that delve into the comings and goings of events on the Sherman Ranch.

UFOS DEJA VU

UFO sightings keep Uintah Basin buzzing

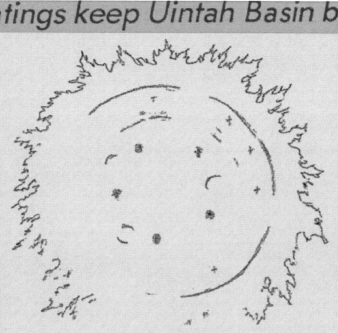

Copyright Deseret News 1978

By Andrea Granum
Deseret News correspondent

ROOSEVELT — The Uintah Basin, already famous for numerous UFO sightings in the past 10 years, has been the stage for a great deal of excitement in recent weeks.

Area residents say that at various locations and at times they have witnessed the flight of an awesome dome-shaped unidentified flying object with intense lights.

Dale Wood, 13, a student at Vernal Junior High School, was the first to see the large silver object, near his grandmother's home about seven miles northeast of Roosevelt. About 10:30 p.m. Aug. 11, while walking to his grandmother's trailer home, he could hear the sound of a finely tuned purring engine, Dale said.

He looked up to the horizon and was surprised to see a silver dome-shaped object. It was surrounded by a very intense green light that was jagged like the flames of a fire, he said.

The object then hovered directly above him and he was able to see the underneath section of the craft. He could see lights there, with the middle light of greatest intensity.

While the craft was overhead, the engine sound was not noticeable, he said. The object remained overhead for a few minutes and, Dale said, he had the impression he was being watched.

Dale said he was very scared, but at the same time he wanted to see what the UFO was going to do. It began to

Wood called the Ute Indian Tribe Police Department in Fort Duchesne. Officer David Murray was dispatched to investigate the sightings. He later said he, too, saw the object near U.S. 40.

The following day, in a conversation with Tehntha Rasmussen, a reporter for the Roosevelt Standard, Mrs. Wood learned that Mrs. Rasmussen and her 10-year-old grandson, David, had also seen a craft near their Ballard home.

While driving home from Roosevelt, Mrs. Rasmussen spotted something out of the corner of her eye. She looked up through the windshield and asked David if the shining object she saw was a plane or something of that nature.

David calmly replied that it was a flying saucer.

At that point, Mrs. Rasmussen, fearing an accident, pulled her car to the roadside and got out to have a better look.

She described the object as being dome-shaped and very shiny. It was moving at a high rate of speed, she said. It headed north to the mountains.

Mrs. Rasmussen and David said they were not afraid of the UFO. She said she had seen a saucer-like object about 20 years ago in the Neola area, so she knew a little about what to expect.

The Woods and Mrs. Rasmussen contacted Junior Hicks, a science teacher at Roosevelt's West Junior High School who has gained some fame in the field of UFO investigation. Since 1965 he has personally investigated more than 400 sightings in the Uintah Basin.

Through Hicks' extensive files, a

The Sherman Ranch may have shapeshifters, but the rest of the state has some pretty stunning UFO reports to explain.

UFO Declassified host Erica Lukes stands at the locked gate of the Skinwalker Ranch and wonders what the heck is going on in there.

UFOS DEJA VU

Right: Though most visitors are locked out, Erica Lukes has been on the ranch and has no doubt but that the stories in circulation are absolutely true.

Below: Erica sports an alien on her arm — but there is no high school prom going on at the Sherman Ranch.

News

The Curse goes back to the connection between the Utah Indians and the land they were forced to leave. According to local legend, the valley was cursed by the natives upon their forced exodus to federal reservation grounds in Utah.

"This land is cursed" says local tribesman

George Horner

Not many people have heard of the curse of the Skinwalker Ranch. Those who dwell there for more than 30 days are subject to "The Curse." "Unless you take some dirt with you, you will never leave this place" says a local tribesman.

Back in the 1890s is when the curse originated, when the town was first settled. Until then it had been home to the Natives who had been living there for decades. In order to settle the land, the settlers pushes out these Indians.

The early settlers possibly felt bad for their actions and therefor started the myth of the curse to ensure that even though they had forced them of their home, they would someday return.

Now the myth of the Curse continues with our current generation. According to locals, this Curse is real. Sally Henricks: lived in Utah for many years before she and her husband left. After seven years they had had enough of the strange happenings and longed to live somewhere free of the curse.

The Curse goes back to the connection between the Utah Indians and the land they were forced to leave. According to local legend, the valley was cursed by the natives upon their forced exodus to federal reservation grounds in Utah.

The legend states, among other things, that no person born in the valley may leave permanently unless a small amount of sand is collected from the river. The sand is supposed to alleviate the curse's effects of a supernatural and metaphysical attraction by the valley's soil to the native individual. *ghorner@mail.com*

A local clipping proclaims: "The land is cursed, says local tribesman."

UFOS DEJA VU

Erica's UFO Declassified show can be heard Friday evenings over KCORradio.com - You can also do a search for her on YouTube.

UFOS DEJA VU

SHAPESHIFTERS, WEREWOLVES, TRICKSTERS, MONSTERS, AND MUCH MORE
By Nick Redfern

PUBLISHER'S NOTE: If there is one individual you would like to go out and share a beer with it would probably be Nick Redfern. I have to hand it to him. I used to be king of the Men In Black story tellers, but he has surpassed me in recent years in collecting reports from our constituents on the topic. I may have taken the only authentic photo of a MIB, but Redfern can rattle off hundreds of real cases, including those of Women In Black. I can't beat him on that aspect. Nick has probably been a guest on our "Exploring the Bizarre" podcast more than any other guest – mainly because he has written more influential books on the paranormal than anyone else, especially since the passing of Brad Steiger.

Nick Redfern is a full-time author and journalist specializing in a wide range of unsolved mysteries, including Bigfoot, the Loch Ness Monster, UFO sightings, government conspiracies, alien abductions and paranormal phenomena. He writes regularly for the London Daily Express newspaper, Fortean Times, Fate, and UFO Magazine. His previous books include Three Men Seeking Monsters, Strange Secrets, Cosmic Crashes, and The FBI Files. Among his many exploits, Redfern has investigated reports of lake monsters in Scotland, vampires in Puerto Rico, werewolves in England, aliens in Mexico, and sea serpents in the United States. Redfern travels and lectures extensively around the world. Originally from England, he currently lives in Dallas, Texas.

Here Nick explains that we have to consider more than just your skinwalker and native witch that has been discovered in the last few chapters, and discover and reflect on the overall circumstances of the "trickster" phenomena, be it werewolves, elementals or your neighborhood variety of Bigfoot. His report that follows will impress this matter upon you.

* * * * * *

While the traditional image of the werewolf is, without a doubt, the first thing that springs to mind when a discussion of shapeshifters takes place, the truth of the matter is that there is a veritable menagerie of such infernal things in our midst.

UFOS DEJA VU

Were-cats, were-tigers, were-hyenas, and were-coyotes are also near the top of the monstrous list. Then, there are the ancient beliefs that those who died violent deaths—or those who were, themselves, murderers—were often destined to return to our plane of existence in the forms of hideous beasts, including wild and savage apelike animals, fearsome black dogs with glowing and blazing red eyes, and mermaid-like things. There are also beings from other worlds: aliens, extraterrestrials, and Men in Black.

Even the legendary monsters of Loch Ness, Scotland, are believed—in certain monster-hunting quarters—to be paranormal beasts that have the ability to alter their appearances at will. As are legendary vampires, who, the old legends suggest, can transform into the likes of bats and wolves.

Collectively, all of these "things" amount to an absolute army of otherworldly creatures and half-human monsters that have plagued and tormented us since the dawning of civilization. And, they show zero signs of slowing down anytime soon. The things you thought were only fit for campfire tales, late-night stories intended to thrill little children, and entertaining monster-movies, are, in actuality, creatures of the real world. Of our world. Shapeshifters are everywhere: they lurk in the shadows, in the deep woods and expansive forests, in dark and dank caves, and in the murky waters of our lakes and rivers. Maybe even, after sunset, in the recesses of your very own backyard, patiently waiting to pounce. And many of them like nothing better than to terrorize and torment us, the human race. Let's take a look at a few examples.

WEREWOLVES

The latter part of the 1800s saw the surfacing of a tale of shapeshifting out of Germany—a country that has a long and checkered history of encounters with werewolves. It's specifically to the town of Ludwigslust we have to turn our attentions, a town with origins that date back to 1724. One particular creature that became almost legendary in 1879 was a large, wild wolf that seemingly was completely unaffected by bullets. The brazen beast would even creep up on hunters and steal their bounty—their dinner, in other words. It's no surprise that word soon got around that maybe the wolf was more than just a nimble animal that had been lucky enough to avoid getting shot. Some thought it was supernatural in nature. Others, in quiet tones, suggested Ludwigslust had its very own werewolf. They were right. Witnesses claimed to see a wizened old witch transform into the deadly beast—and back again.

In April 2016, a very strange story surfaced out of the north of England. And to the extent that not just the local media, but the national media, too, were busy chasing down the strange and sinister story of what has become known as the "Werewolf of Hull," reportedly an eight-foot-tall, hair-covered monster. The case was, however, notable for the fact that several of the witnesses claimed the beast shape-shifted from a terrible, foul monster into the form of a black-cloaked old witch.

Most of the reports surfaced in and around the vicinity of what is called the Beverley and Barmston Drain, a land drainage operation, the origins of which date back to the latter part of the 1800s. A tunnel that carries the drain can be found below an old bridge on Beverley Beck, a canal in East Riding, Yorkshire, England—a location where a number of the encounters with the hair-covered thing have taken place.

So, what might the creature have been? The UK's media picked up—and picked up quickly and widely—on a local legend of an abominable beast known as "Old Stinker." It's a terrifying half-human/half-animal thing that has a long history in the area, a history that dates back centuries. So the story goes, the name came from the legend that the creature allegedly suffered from severely bad breath! And its physical appearance was said to have been no better either: it was covered in hair, and had piercing and glowing red eyes. On top of that, it regularly devoured the corpses of the recently deceased.

WERE-CATS

Jennifer's encounter occurred in her small Oregon hometown in the fall of 2011. Late one night, Jennifer saw what she assumed was a large black dog walking towards her, on the sidewalk. As the creature got closer, Jennifer was horrified to see it was actually a significantly-sized cat. As in the size of a jaguar. Jennifer was about to make a run for cover—but to where, at that time of night, she admitted she had no idea—when the cat suddenly stood upright, changed its appearance and ran across the road, at a phenomenal speed, and vanished into an alley on the other side of the road. Notably, Jennifer said that the cat did not lose its black color as it changed into human form, nor did it lose its catlike head. It was, according to Jennifer, something that appeared half-human and half-cat. A definitive were-cat.

ELEMENTALS

Of the many and varied kinds of "elementals" that were said to possess the awesome powers of shapeshifting, certainly one of the most mysterious, and strangest of all, was the Dryad. It was a definitive entity of magical proportions that took shapeshifting to a truly unique level. The Dryad was a supernatural entity that features heavily in ancient Greek mythology, and which was exclusively associated with forests, woods, and trees. There was a very good reason for that: the Dryad had the uncanny and eerie ability to transform itself into a tree!

COYOTES

Perceived almost unanimously by Native Americans as a trickster-like animal, the coyote is said to have the ability to control the weather, specifically rain and storms. And, like so many other trickster entities—such as fairies and goblins—the coyote can be friendly, playful, and helpful. But, and also like all tricksters, the coyote has a dark side: it can be manipulative, deceitful, and even deadly, and as the mood takes it. As for what Native American lore says of the shapeshifting abilities of the coyote, we are told that the animal can take on human form, usually in the guise of a man with a large mustache. Coyotes are said to be able to transform into the forms of birds, fish, and cats. Also according to Native American mythology, witches and those familiar with magical rituals can transform themselves into coyotes. Thus, a coyote seen running wildly late at night may well be a shapeshifting witch or wizard, embarking on some dark and disturbing mission.

THE KITSUNE

Japan has its very own shapeshifter: the Kitsune. It is a word that means "fox." Japanese lore has long maintained that each and every fox that lives—and that has ever lived—has the ability to take on the appearance of a human, whether that of a man or of a woman. And, like many of the shapeshifters that we have focused on so far, the Kitsune alternates from being a malevolent creature to a placid and helpful entity. Mostly, however, it acts as a classic Trickster – manipulating people, and playing endless mind-games.

And that's just the start of things: the Men in Black who terrorize UFO witnesses have been seen to morph into the forms of blazing-eyed, black dogs. The legendary Mothman of Point Pleasant, West Virginia has been described variously as a winged humanoid, as giant bat-like animal, and as a huge bird. Clearly, the beast is able to take on multiple guises. Bigfoot witnesses tell of the creatures changing into bright balls of light. The legendary creatures of Loch Ness are said to have the ability to take on the forms of beautiful women and large black horses.

If you think that shapeshifters are merely the things of folklore, mythology and legend, it's time to think again. Shapeshifters are here, there, and everywhere—and in multiple, monstrous forms, too.

Another fabulous work by Fortean researcher, Nick Redfern, Shapeshifters. Morphing Monsters, and Changing Cryptios is a must read. It was published in 2017 by Llewellyn.

UFOS DEJA VU

Left: Nick hangs out with our homies Allan Benz and Charla Gene.

Below: A photo Nick is proud of. Standing next to Roy Thinnes, star of the legendary TV series, "The Invaders."

UFOS DEJA VU

Carol Ann Rodriguez depicts a very Were-Cat.
Even Nick would have to admit that most don't fly. Though don't tell John Keel that. He discovered the Mothman creature while trying to track down Tom The Flying Cat in West Virginia.

We all know by now
when
the sign says
KEEP OUT
they mean
KEEP OUT!

SECTION FOUR
DOWN IN THE VALLEY

UFOS DEJA VU

UFOS DEJA VU

THE SAN LUIS VALLEY – MORE THAN YOUR AVERAGE HOTSPOT
By Sean Casteel

PUBLISHER'S NOTE: As "coincidence" has it, I knew Chris O'Brien before logic tells me I should have known Chris O'Brien. Back in the notorious rock and roll 1970s I was booking and producing some local NYC bands. I actually can say I had Satan under my management. No not the guy with the horns and tail, but a fire breathing singer born Isreal Jones from New Orleans (i.e., friend of Dr. John the Night Tripper and Black Oak Arkansas). Just think Gene Simmons from Kiss and you will get the idea. In fact, Satan always said that he was first with the fire eating and that Kiss had copied his routine. I don't find this hard to believe as Satan and Simmons played the small venues in those days and even rehearsed at the same studio, Talent Recon. I was also buddies with drummer Bleu Ocean, a Native American who was born on Mount Shasta. Bleu assisted at our conferences as our audio visual liaison with the speakers since I barely know how to plug in something. He also organized a drum solo for Pink Floyd with like 50 drummers, for which he became quite famous. Chris O'Brien, in addition to being an established author, radio host and researcher of all things "dark and dreary," happens to have played in bands for decades, and we knew him from the Record Plant and other recording studios as a fine musician. We didn't exactly chum around with him but we knew who he was and respected his artistic talents. He has built quite a reputation as a keyboard player, music producer, digital artist and videographer. Because of what would seem to be "big city trappings," we often wondered why Chris left the Big Apple for the wilds of Colorado and the San Luis Valley in particular. To find out why check out an interview we did with O'Brien on "Mr. UFOs Secret Files" YouTube — www.youtube.com/watch?v=FDOZI8G3vN0&t=474s

* * * * *

But its time to climb in the saddle and let Sean drive the herd— what is left of it after the mutilations which have been on going in the Valley since Snippy went down not of his own accord.

* * * * *

There are few UFO sightings hotspots in the United States—or the world, for that matter—that can rival the Greater San Luis Valley. The valley straddles Colorado and New Mexico, occupying 13 counties within the two states.

The sightings are so frequent that a transplanted resident named Judy Messoline built something called "The UFO Watchtower," an observation platform, campground and gift shop with a 360-degree view of the San Luis Valley. The tower is located in Hooper, Colorado.

A website called Roadside America tells the following story: Messoline had moved to the area to raise cattle, an effort she soon abandoned due to the harsh conditions and the lack of adequate grazing land. Her neighbors had mentioned seeing strange things in the night sky of the Valley, and she knew that UFO watchers often visited her ranch after dark. To earn some money after selling her cattle, she opened the campground, built a small saucer-shaped dome as a gift shop and surrounded it with a ten-foot-high viewing platform, not exactly an extremely tall structure but nonetheless... .

"Unapologetic," the website says, "she called it 'The UFO Watchtower' because, as she told us, 'When you're already at 7,600 feet, you don't need to be much higher.'"

The tower opened in the summer of 2000, and Messoline didn't expect that anyone would see anything from it, other than the distant mountain ranges and the stars at night. From the eastern view of the tower, one can see the Great Sand Dune, the largest sand dune in the world at 75 feet high.

But what Messoline didn't know was that the San Luis Valley is revered among flying saucer buffs as one of the best places in the world to see UFOs. And her watchtower is right in the middle of it.

Another transplant to the San Luis Valley is researcher and author Christopher O'Brien, about whom there is more to come later. O'Brien has kept meticulous records of sightings in the area.

As an example, O'Brien writes, "I received the following report from a reliable skywatcher who lives just south of the Colorado/New Mexico border, in the foothills 1000 feet above the east side of the San Luis Valley floor."

*** The woman was a resident of Taos, New Mexico, who was visiting friends just south of Questa, New Mexico. The woman inadvertently shot some late afternoon video footage of Ute Mountain, about 15 miles west of her location across the Rio Grande Gorge. She captured a formation of nine anomalous objects that she hadn't noticed while filming. Later, at home, when she viewed the footage, she noticed something streak across the screen at extremely high speed.

*** In Chafee County, Colorado, a man, his fourteen-year-old daughter and her cousin, also fourteen, sighted a cigar-shaped object a long distance away, 40 to 50 miles over the Presidential Mountains west of Buena Vista. The object stayed at a constant 30 degrees above the horizon. There was no movement and no vapor trail. In the two to three minutes of videotape footage taken at the scene, other objects appeared that looked like two barbells connecting the main object. The object dimmed out slowly, in about 15 seconds, in the area of the sky that a jet seemed to pass through a minute later.

*** Another local woman reported watching an iridescent large green ball that hovered low in the sky over the Center, Colorado area before slowly drifting to the western edge of the San Luis Valley. She described the object as a lot larger than a full moon and located approximately 25 degrees above the horizon. She estimated that she watched it for three to four minutes before it seemed to slowly turn a pale yellow and fade out.

*** Two witnesses observed three to four lights hovering over the Sangres Mountains. The planet-sized lights appeared to be pulsing at each other and alternated between white, red and green. The event lasted around four to five minutes before the lights descended out of sight behind the mountains. A week later, the same two witnesses saw three unusual maneuvering lights above the Sangres for several minutes. One light was described as larger than the other two.

Read on to learn more about Christopher O'Brien and the results of the more than 20 years he has spent cataloging the seemingly unending flow of paranormal events in the San Luis Valley. From UFOs to Bigfoot, the Valley has been host to a huge number of events seen by a large swathe of the local population, including the indigenous Native Americans, who have spoken of the alien presence there for centuries.

THE SAN LUIS VALLEY –
HOME TO CREATURES HUMAN AND OTHERWISE

When investigator and researcher Christopher O'Brien first moved to the San Luis Valley in Colorado in 1989, he didn't have a clue as to what lay ahead of him.

"I've always had an interest in the paranormal," O'Brien said, "the things that aren't really supposed to happen, the things that they don't teach you about in school. And I'd known just peripherally about stuff that had been going on here over the years. I'd heard stories from books that I'd read and, after I moved out here, from people that I met here.

"But I had no idea," he continued, "that this place was like a paranormal Disneyland. In about 1992, all of a sudden a lot of very strange things started happening."

O'Brien began to investigate the odd happenings, which included cattle mutilations, black helicopters and UFO sightings galore. He also interviewed many witnesses to the events taking place there, initially to write articles for the local newspaper.

"Within three months," O'Brien said, "I was on national television. So here I am. I've written a couple of books, written a lot of magazine articles, and been on

UFOS DEJA VU

all kinds of television segments. And I'm probably more confused now than when I started."

O'Brien's first book was called "The Mysterious Valley" (St. Martin's Press, 1996).

"It's sort of like an inside view of not only just an amazing time period here," he said, "from 1992 to 1995, when there were just all kinds of very, very strange events that took place, but it was also [written] as a way for people to kind of understand what I went through, immersing myself into this whole area."

The follow-up was called "Enter The Valley" (St. Martin's Press, 1999).

"The second book has a lot more historical facts," O'Brien explained, "about what has gone on here over the past several hundred years. And then the third book will continue on from there and have a lot more analysis and insight based on [my] ten or eleven years of investigative work."

The historical research O'Brien refers to turned up some fascinating characters from Colorado's past. One of them, a would-be gold miner turned cannibal, surely qualifies as a creature of the human kind.

"Most people realize," O'Brien began, "that Colorado was first explored and the first real settlers moved here as a result of gold and other precious metals being found here. And Alferd Packer was like many people who arrived here from elsewhere. He was interested in finding gold and striking it rich. He and a group of men arrived on the western slope of Colorado in the early 1850s, late in the season."

Packer and his associates were trying to get to the central part of the state in the dead of winter, where temperatures can drop to 50 degrees below zero and snow falls for days and days at a time. Packer and his team were even told by local Indians that it was too late for them to make the journey.

"Well, they waited around and then decided 'What the heck, we're going to be the first ones there,'" O'Brien continued. "So instead of waiting until the real spring thaw, they started out to try to get to the area of the gold fields, which is just west here of the San Luis Valley."

There were two possible routes. One meant following the Gunnison River, which was the longer but safer way to go. The other route involved going over the mountains.

"They only had a few days' worth of rations with them," O'Brien went on. "If the weather had been perfect, if it had been the summertime, they would have had enough rations to get there from where they had been holed up for the winter."

Packer and his cohorts tried first one route, then abandoned it for the other, both times ending up lost amid 12 to 15 foot snowdrifts. Things went from bad to worse.

UFOS DEJA VU

"They ran out of food," O'Brien said. "The story gets a little controversial at this point, but what we do know is that about three months later, Alferd Packer shows up in Saguache, Colorado, which is the county seat of the county I live in here, and he claims he was the only survivor of the group of six guys."

Which immediately aroused the suspicion of the local sheriff, whose fears were confirmed the following June when a group of tourists and writers for the magazine "Harper's Weekly" reported that they had found the graves of Packer's victims.

"It was very obvious that the guys had been butchered," O'Brien said. "Literally."

Packer finally admitted that he had cannibalized his buddies, though his story of exactly how it all happened changed many times. Famous at the time as "The Colorado Cannibal," he was tried and convicted of all five murders and sentenced to hang. Four days before his scheduled execution, his two attorneys managed to obtain a stay based on a technicality. Colorado had no death penalty provision for a capital crime committed before Colorado became a state, according to O'Brien. At his second trial, Packer was again found guilty and sentenced to 40 years in the Canon City Penitentiary.

Polly Pry, a sympathetic journalist from "The Denver Post," later helped Packer obtain a pardon, and in a thank you note to her he wrote, "I have never closed my eyes in sleep since without that ghastly vision of the smoldering camp fire, the dead companions and the lofty pines drooping with their weight of snow, as if keeping a sorrowful death watch. But, those who have never been without their three meals a day do not know how to pity me."

Packer died of a stroke in 1907, a broken man.

O'Brien also uncovered the story of the first serial killer in the United States, Felipe Espinoza of the San Luis Valley.

"I tried my darnedest to find earlier examples of a serial killer in the United States," he said. "Generally, the whole idea of a serial killer really became popularized with Jack the Ripper in England. But very few people know about the Espinozas, even here in Colorado."

Felipe Espinoza was a native Mexican who lived in Vera Cruz as a little boy.

"During the Mexican-American War," O'Brien said, "when Admiral Whitfield Scott was bombarding Vera Cruz from the harbor with American warships, one of the shell blasts took out Espinoza's whole family. He was orphaned basically. His only living relatives at the time lived up in Colorado, in Conejos County, which is the county that shadows the New Mexico/Colorado border in the San Luis Valley here.

"So he was sent up here," O'Brien went on. "Back then, there were mostly Spanish-American settlers in the southern part of the valley. One night, as the story goes, he had this vision or dream in which the Mother Mary came to him and told him he had to avenge the deaths of his six family members. And the way he was told by the Virgin Mary to avenge his family's deaths was to kill a hundred gringos or a hundred Americans for each one of his family members that had died in the shell blasts years before."

Espinoza managed to enlist his uncle and/or cousins (the history is confused on that point) to join his crusade and he began a three or four month rampage of bushwhacking Americans.

"By the time that it was over," O'Brien said, "they had bushwhacked between thirty-five and forty people. What made it even stranger was that they would blacken their faces, and when they would bushwhack these cowboys or settlers or miners, Espinoza would take an ax and hack open their chests and pull their hearts out. There's some question as to why he did this and what exactly they did with the hearts."

After the first dozen or so murders, the Territorial Governor sent out a posse and offered rewards for the arrest and capture dead or alive of the family. Their efforts proved fruitless for a couple of months, and to aggravate matters further, innocent people were mistakenly accused of being the Espinozas and hung on the spot.

"So it was getting really ugly here, and it was quite a sensational news story at the time. People were traveling around in armed groups because they were so afraid of the Espinozas."

The Territorial Governor finally called upon famous mountain man Thomas Tobin, a contemporary of Kit Carson and Jim Bridger. Tobin succeeded in finding the murderous family. He killed them and cut off their heads. He was returning to collect his reward, bearing the severed heads in a burlap bag, when he was swept away by a river current and lost two of the heads, hence the confusion as to exactly who Espinoza's accomplices had been. Tobin never received his $500 reward, however, because there were no public funds available at the time.

"Legend has it that Felipe's head ended up in a pickle jar in the sideshow of a circus," O'Brien said. "So there's America's first serial killer. Not many people are even aware of him."

According to O'Brien, something similar to Espinoza's extreme brand of Catholic fundamentalism is still practiced quietly there in the San Luis Valley. The sect's name in English is "The Brotherhood of Blood." Their rituals include stripping naked and flagellating themselves with a cat o' nine tails and marching in long processions over cactus fields barefoot.

"Back in the old days, they used to take the most worthy, pious brother and crucify him on Good Friday," O'Brien said.

The site of the crucifixion would be marked by a white cross, many of which are still visible on the hilltops in northern New Mexico and southern Colorado.

"Now I think it's pretty rare that they actually crucify somebody. It's more of a mock crucifixion now. But the penitents are alive and well. They still are very secret but very real – the most fundamentalist of all Christian sects that I've been able to find. It kind of puts a new twist on Christianity."

The sister of a friend of O'Brien's married a member of the sect, who would occasionally show up at the friend's home to visit.

"He would sit back rather gingerly in his chair because his back was all flayed from doing these rituals," O'Brien said. "So it's the real deal."

Meanwhile, the San Luis Valley is also home to more conventional "creature" appearances.

"We've had quite a number of Bigfoot sightings that I've been investigating in this part of the country," O'Brien said. "We have, right on the border of the San Luis Valley, these two very large mountains that sit separate from one another and separate from any other surrounding mountains. The locals have a lore that says the Bigfoot lived inside the mountains, and the Pueblo Indians believe that's where the Creator lives and that he appears to mortal men in the guise of a Bigfoot."

Local law enforcement officials in Conejos County recently videotaped some sets of tracks that wound through and over a variety of terrain for hundreds and hundreds of yards, O'Brien said. Seven separate Bigfoot sightings occurred that same week.

Reports have come from the eastern side of the San Luis Valley of something called a Thunderbird, a giant bird that looks similar to a raven or a raptor, but with a 40-foot wingspan. The bird is said to soar quite high and create a visual effect that looks something like a hole in the sky itself.

A serpent deity called Talulukang is rumored to haunt the Valley as well.

"He's the serpent underneath the mountain holding the world in check," O'Brien explained. "It's very similar to the Tibetan tradition of the Nagas, which are said to be large serpent-like entities that live in the Himalayas and have the exact same responsibility. The San Luis Valley is one of the places where Talulukang is supposed to be. The shaman-medicine men-type are able to see this creature, but it's a creature that is normally underground and he doesn't reveal himself very often."

The leader of one of the first major Spanish incursions into the valley, prior to President Thomas Jefferson's Louisiana Purchase, reported in his diary that he and his troops saw weird flying lights at night in the mountains and also heard strange very low humming sounds coming from beneath the mountains. What the Conquistadors experienced then still goes on today, according to O'Brien.

The San Luis Valley is also known as a hot spot of concentrated cattle mutilation activity.

UFOS DEJA VU

"I've always wondered," O'Brien said, "if the paranormal aspect of that whole thing may be some sort of predator that's going around – that we may be dealing with some sort of 'dimensional' or just 'otherworldly' type predator. But again, that's a couple of books down the road."

Perhaps O'Brien is quite right when he calls the San Luis Valley a "paranormal Disneyland." For the moment at least, he offered the following in conclusion:

"There's not many places in the country that have the variety and intensity of 'creatures and features' that this place does. I think I've established that pretty accurately in more than 20 years of investigative work. So let your readers decide if that's an accurate statement or not."

www.OurStrangePlanet.com

SUGGESTED READING, BOOKS BY CHRISTOPHER O'BRIEN:
THE MYSTERIOUS VALLEY
ENTER THE VALLEY
SECRETS OF THE MYSTERIOUS VALLEY
STALKING THE HERD
STALKING THE TRICKSTERS

O'Brien's well researched book caused a stir when first published.

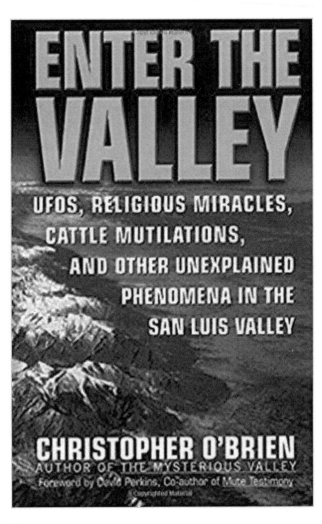

142

The Pueblo Chieftain

96th Year · Pueblo, Colorado, Sunday, September

SIGHTINGS REPORTED

Dead Horse Riddle Sparks UFO Buffs

The bones of Snippy were recently offered for sale on ebay. More morbid than the mutilation itself!

By THE CHIEFTAIN STAFF

ALAMOSA (C-SJ)—The death of a horse in the San Luis Valley during early September appeared Friday to have sparked a renewed interest in unidentified flying objects throughout the West.

In Houston, Tex., a group of UFO enthusiasts left Friday night for the San Luis Valley to inspect the dead horse and also to investigate many reported UFO sightings in the valley.

A Texas radio station reported a flood of calls Friday morning after broadcasting the story of the bizarre death of the Appaloosa owned by Mr. and Mrs. Berle Lewis of Alamosa.

The horse was found dead Sept. 9. The flesh and hide were stripped from the neck and skull leaving the remainder of the body intact. High levels of radiation also were reported in the area of the animal's body.

See "Something"

Lewis, the owner of the horse, said that increased sightings of UFOs in the area have been noted during the past six weeks. "We see something — I won't say what it is—every night."

In the Houston area at 6:08 p.m. Friday, UFO seekers said they spotted two high-flying cigar-shaped objects — each about half the size of a football field—on a course "plotted" to take them over the Southern Colorado area.

In Boulder, James Wadsworth, research investigator for the University of Colorado UFO project, said an investigator will go to the San Luis Valley this weekend to investigate the horse's death.

Mrs. Charles Blundell, wife of the maintenance crew foreman at the Great Sand Dunes National Monument, Friday said the latest sightings brought back to her mind a curious incident which happened to her in late 1966.

Mrs. Blundell said she saw a crescent-shaped object over the Sand Dunes last December. Later she painted a picture of what she had seen.

"Peculiar Man"

During the past summer, a "peculiar man" approached her and asked to buy the painting. She said she did not want to sell, but did set an exorbitant price on the painting. She said the man, who said he was "not of your world," told her he would return in October to buy the painting. He has not returned.

The UFO seen by Mrs. Blundell glowed brightly. It also was seen by her daughter, Terri, 14, and neighbors. Mrs. Blundell says she does not believe in flying saucers and believes there is a logical explanation.

Similar Sighting

A week ago, scientists at the National Atmospheric Research Center near Palestine, Tex., reported seeing a crescent-shaped object in the sky, much like the object described by Mrs. Blundell.

Texas observers also have reported a sharp increase in sightings of UFOs in the area during the past two weeks.

One thing is certain in Southern Colorado. Saucers may or may not exist, but people are seeing something. Snippy the Appaloosa is d e a d, and the circumstances surrounding his death are far from clear.

Camera picks up strange object "floating" across the San Luis Valley.

UFOS DEJA VU

Home on the range. Chris examines a recently mutilated animal in the Valley.

Snippy the horse made history when he became the first nationally recognized mutilation of an animal.

Tim Beckley visits with Chris, taking time to promote their individual works of art.
Photo by Charla Gene.

UFOS DEJA VU

Would be sky watchers gather at the UFO Watch Tower in the Valley hoping
for a sighting of their own. They are often rewarded with one!

ever, the actual location where they experience these events remains important as well.

In addition to so-called paranormal events, hotspot areas of the world also appear to feature unusual geophysical properties that may account for their higher than normal incidences of unusual events. These geo-energetic elements, when further defined and studied, could potentially provide academic motivation to investigate the full width and breadth of these hotspot regions' unexplained activity. Without question, there appears to be an emerging public interest in these specific locales (and the events they feature) and word about the documented activity has been slowly filtering out into the culture at large.

Hotspot regions are prime territory for explorers interested in all kinds of unusual phenomena, including portal site visits and/or investigations. However, many of these areas are sacred sites and their mythic traditions should not be treated lightly by casual travelers.

I spent ten solid years from 1992 to 2002 investigating and documenting events that occurred in and around the SLV. I uncovered thousands of stories from the recent past and hundreds of current reports of the unexplained; many of these gems now encrust the extensive database I developed covering this crown jewel of paranormal hotspots. Something very intriguing is going on in the SLV and in other hotspot regions, something well worth a world explorer's interest and study—and we may be at a time where these areas are finally awarded the attention they deserve.

NIDS

One high-profile scientific effort is worthy of mention. For over ten years the National Institute of Discovery Sciences (NIDS) conducted a monitoring program at the infamous Sherman Ranch "portal site" in Utah's Uintah Basin. The book Hunt for the Skinwalker (Paraview Books, 2005) by Colm Kelleher and George Knapp provides a somewhat incomplete yet recommended recounting of the investigation. Given rumors and circumstantial evidence of its possible involvement in this case, it is not outside the realm of reason to suspect that the US government has been monitoring other hotspot areas (e.g., in the SLV and the Uintah Basin) for decades. I think it's not-so-safe to say there is probably more covert action going on behind the scenes in these regions than investigators and the media are acknowledging.

UFOS DEJA VU

So, has the government military-industrial complex already forged ahead, decades ago, in an effort to define and study the enigmatic physical properties that appear to be at work in these hotspot locales? Going even further, is it possible that they may have come to a new understanding of these scientific riddles and are actively (with the aid of high technology) utilizing these regions' unique energetic properties for unknown purposes? These questions have not been adequately addressed by the paranormal research and investigation crowd, which seems all too preoccupied with rehashing Roswell and other probable red-herring cases. I have become suspicious of this hidden potential element because I have logged hundreds of reports that suggest apparent military involvement during incidents featuring extraordinary circumstances, and events beyond simple mundane understanding.

Complicating the recipe further, many North American hotspot regions (especially those found in the Four Corners area of the Southwest) have a nearby governmental/military presence that appears to be interested in expanding its control further into these specific areas. It seems to this investigator that many reports of unusual aerial phenomena in hotspot regions actually identify military activity. Your tax dollars at work!

This apparent military presence is especially prominent in and around the San Luis Valley, where I have actively investigated. This presence has muddied the investigative waters numerous times, rendering the task of defining unusual incidents even more difficult. Proto-scientific and amateur subcultural examination of these traditions may help explain the true nature of what is manifesting today in remote areas like the SLV, but how does anyone investigate and explore things so timeless and inexplicable without a budget or a clue? I can speak to this question, because that is exactly what I did.

When I casually began my amateur investigation of the SLV in 1990, I didn't have a clue what was truly occurring, or what I was getting into, but over time (and with the help of several key experts and a lot of research) I was able to devise a fairly effective approach to documenting these events, as detailed in my Mysterious Valley books. At first I innocently assumed all UFO sightings involved "aliens" piloting non-conventional craft, but my thinking on this and other phenomena quickly succumbed to objective analysis, and the factoring out of any mundane explanations.

A Quick Historical Overview of the San Luis Valley

A perfect example of a so-called hotspot region is the San Luis Valley, the largest alpine valley in the world. It is also the largest and highest freshwater aquifer in North America. Teams from the Smithsonian Institution and other scientists have been gathering evidence of human habitation at various sites in the valley for decades, and it is established that this sacred place has been visited by humans for almost eleven thousand years.

US Army Lieutenant Zebulon Pike led the first acknowledged American party over the Sangre de Cristo Mountains, down into the world's most mysterious sand dune field, and then out onto the valley floor in 1807. He had been commissioned to explore as far west as the Arkansas and Red Rivers to ascertain the extent of the Louisiana Purchase. His journal entry on January 28th, 1807, stated the following:

After marching some miles, we discovered...at the foot of the White Mountains [today's Sangre de Cristos] which we were then descending, sandy hills...When we encamped, I ascended one of the largest hills of sand, and with my [spy]glass could discover a large river [the Rio Grande]... The Sand-hills extended up and down the foot of the White Mountains about 15 miles, and appeared to be about [five] miles in width. Their appearance was exactly that of a sea in storm, except as to color...

Below the dunes, the 4,000 square mile, semiarid desert SLV floor is perched at an average elevation of 7,600 feet, over a mile-and-a-half above sea level. The vast valley floor averages less than five inches of rainfall per year. The valley's entire wishbone shape, over 130 miles long, is ringed by majestic forested mountains that soar skyward on all sides.

On the eastern side of the middle part of the SLV, the Great Sand Dunes afford a commanding view of the valley. Behind the dunes to the east stands a solid wall of rock soaring to heights of over 14,000 feet, the imposing Sangre de Cristo Mountains (or Sisniijiini in the native Dine' tradition). This cluster of promontories at the valley's midpoint, which I'll call the Blanca Massif for convenience, contains the valley's highest mountains. This 25 square-mile jumble of peaks stands like a host of brooding sentinels in the dawn. For thousands of years, Blanca Peak and its lofty neighbors have been the focus of Native American myth and tradition. This stretch of the Sangres is rumored to contain doorways or portals to another realm where "all thought originates." Indeed, the area of the Sangres around the dunes appears to be ground zero in the mysterious SLV. The Sangres are the longest continuous mountain chain in North America. The southern end of the valley is dominated on its eastern side by the Sangres' Wheeler Peak, the highest mountain in New Mexico.

The second youngest mountain range in the continental United States, the Sangres owe their jagged appearance to their age of less than a million years. The incredible view of the surrounding mountains from Highway 17 prompted a group of visiting Tibetan monks to call the San Luis Valley "America's Tibet."

The entire western side of the SLV is bordered by the older San Juan Mountains, which rise above a labyrinth of deep valleys and roaring rivers to their west. The famous Rio Grande River originates in the San Juan Mountains, just west of the SLV near Creede, Colorado. From there, it snakes its way into the valley's mid-point and then heads southward to the Gulf of Mexico. The Sangre de Cristo/Rio Grande Rift is considered to be the second longest rift valley on the planet, bested only by the Great Rift Valley in Africa. Straddling the backbone of North America, the San Juan and Sangre de Cristo Mountains merge at the extreme northern end of the valley in Saguache County, after emptying over one hundred creeks into the largest freshwater aquifer in the United States.

The southern portion of the valley was the first officially settled region of Colorado, and two-thirds of the way down the valley's length, below the Colorado/New Mexico border, are some of the oldest European settlements in the United States. San Luis, in Costilla County Colorado, founded in 1851, is the oldest town in Colorado. As a result of the sixteenth and seventeenth century influx of settlers migrating north from Taos and the New Mexico Territory, today the southern part of the geographic valley has a rich cultural tradition with Spanish-speaking residents making up half the population. Physically and metaphorically isolated from the outside world, this unique Hispano subculture has developed its own special character, combining mystical elements of indigenous Native American beliefs with an Old World style of fundamentalist Catholic piety. This close-knit Hispanic population is very superstitious, wary of outsiders, and their peculiar beliefs incorporate much myth and magic.

Hotspot regions like the San Luis Valley are veritable magnets for reports of paranormal events. But unfortunately, any attempt to define what constitutes a truly "paranormal" event is wrought with perilous philosophic and scientific challenges, and a lack of hard data. Prior to my arrival, knowledge of past-unexplained activ-

UFOS DEJA VU

ity had not traveled out into the mainstream of the valley culture. As in prehistoric times, experiencers' and witnesses' quiet descriptions to friends and family filtered only slowly into the greater local population, with the details subtly shaded or lost as the stories were told and retold. In the past 100 years, small town papers occasionally hinted that these unexplained events were occurring only if reporters were assigned to the story. More widely-known accounts and personal experiences were recounted at family gatherings, picnics, at the post office, over

the backyard fence and out in the grocery store parking lot. Knowledge of these events slowly disseminated around the community. But in this modern age of instant communication, local knowledge of inexplicable events echo around the SLV. And, as in other hotspot regions, word of unexplained activity appears to be firmly embedded in the population.

With the advent of the popular 1992 television program The X-Files, these formerly socially taboo subjects cemented themselves inside America's collective psyche. Even though the stigma attached to these subjects has softened, many people immersed in Western culture still find these subject matters unsettling, and the end-of-broadcast Twilight Zone theme ("de-de-de-dooo") and "little green men" jokes continue to litter local TV news coverage of the inexplicably sublime. The father of "gonzo" journalism, Hunter Thompson, used to assert, "when the going gets weird the weird turn pro"—an axiom that could be applied to this particular paranormal investigator.

When I moved to the mysterious San Luis Valley in 1989 little did I realize that I would spend the better part of the next fifteen years investigating, researching and documenting around a thousand unusual events—all occurring within the well-defined confines of this specific area. Having experienced them as well, I definitely resemble a Thompsonesque gonzoid investigator of the weird.

Looking over my database chronicling these events, I suppose you could say this approach worked in the SLV. If this is so, it should also work elsewhere. The key was coordinating efforts and communicating with an assortment of local law enforcement officials, a skywatcher network, other amateur and professional investigators and local newspaper reporters. Utilizing these techniques, SLV residents investigated an intensive seven-year wave of unexplained phenomenal events between 1992 and 1999. As a result of this effort, this forgotten region at the

UFOS DEJA VU

top of North America was has been determined to be the world's Number One per capita UFO hotspot by the Computer UFO Network, with 257 sightings per 10,000 in population; it could be called America's most active mysterious paranormal hotspot. Why are craft buzzing around up here in the rare southern Colorado air? Why are people seeing Bigfoot and choppers and fireballs and ghosts and elementals and other assorted weirdness?

Native American Traditions in the SLV

In the process of familiarizing myself with this "mysterious valley" Petri dish and its prehistoric tradition, I could not overlook the thirteen different tribes of Native Americans who are known to have visited here over hundreds, and some, probably thousands of years. Let's start at the beginning.

Some members of these tribes still regard the SLV and its surrounding mountains with a mystic reverence. Upon researching these special areas, one gets the sense that there is a connection between current unexplained events and an ancient tradition of sacredness, or specialness, that extends back in time to the first experience in the area. In the case of the SLV, the Crestone-Navajo custodian of Sisnaajini (or the Black Sash Medicine Belt of the Sangre de Cristo Mountains) told me that the portion of the Sangres range that extends from the Blanca Massif north to Crestone plays an important role in the mythic tradition of many Southwestern, Great Basin and Plains Indian peoples.

The Dine', or Navajo, were relative latecomers to the area, but their creation myth (in which the people, who have been surviving underground since their fourth world was destroyed, now emerge into the fifth world) attaches particular significance to the San Luis Valley. The following passage was excerpted from anthropologist Peter Gold's groundbreaking 1990 book, Navajo and Tibetan Sacred Wisdom: Circle of the Spirit:

Let's first consider the most important of the four sacred mountains, Blanca Peak (Sisnaajini), or East Mountain. East Mountain is a distinctive, snowcapped, pine and fir-clothed peak in the Southern Rockies of Colorado. It is considered the 'leader mountain,' because it stands as the holy mountain of the east, the place

154

of beginnings, the dawn. It is associated with the guiding light of the day and the qualities that dawn universally [invoked for] the first people to emerge into the fifth world... by a bolt of lightning—a 'thunderbolt'—whose intense light and quality of energy is most appropriate to that of the dawn...

In Hunt for the Skinwalker, the aforementioned book on the Uintah Basin "Sherman Ranch" hotspot case, authors Colm Kelleher and George Knapp acknowledge the SLV as a location worthy of further sociological research:

Like many other tribes and bands, the Navaho visited, hunted in, and inhabited the San Luis Valley, off and on for hundreds of years. Historians believe that the Navaho were finally ousted from the valley by...the Utes. It is a development the Navaho people are not likely to forget, since they regard the valley as a special place and a fundamental cornerstone of their culture. Mount Blanca [sic], the fourteen- thousand-foot peak that towers over the valley, known to the Navaho as Tsisnaasijini' the Sacred Mountain of the East, is revered as one of four mountains chosen by the Creator as a boundary for the Navaho world. It is considered to be an essential component in the Navaho quest to live in harmony and balance with both nature and the Creator. If the Navaho were Christians, Mount Blanca would be their Bethlehem. If they were Jewish, it might be their 'Wailing Wall.'

Several indigenous peoples have a sacred tradition relating to the purported location of the Sipapu, the "place of emergence" into this current world. Interestingly, as in the Dine' tradition, this entrance may be in the San Luis Valley. The exact location of the Sipapu may never be known by nonnatives, but Dollar Lake and Head Lake in the

San Luis Lakes State Park are two of several possible locations. Another tradition mentions Winchell Lake, located high on the Blanca Massif, as being the location of the Sipapu. Sources observe that the Jicarilla people revere Silver Mountain located south of La Veta Pass. Kelleher and Knapp further observed in Hunt for the Skinwalker:

Not surprisingly, the [San Luis Valley] region also oozes Native American mysticism and legend. The Yuma culture was in the valley five thousand years before the birth of Christ. The list of tribes, bands, and peoples that are known to have moved in and out since then is long. Among those indigenous groups that managed to survive into this century, the San Luis Valley is almost universally revered as a special, mystical place. The Tewa Indians, descended from the Pueblo people and now living in New Mexico, believe that the San Luis Valley is the equivalent of the Garden of Eden. The Tewas say the first humans to enter this world crawled up through hole in the ground to escape their previous plane of existence. Native Americans who live in the valley today say they were taught that the creator still lives in the mountains that surround San Luis and that He sometimes appears to humans in the form of a Sasquatch...

San Luis Valley Crypto-Creatures

The epicenter of Bigfoot sightings in the state of Colorado is along the Colorado-New Mexico border, near San Antonio Peak, said by the Tewa to be home of the god of the underworld, Maasaaw.

The following report was filed in the Bigfoot Research Organization database http://www.bfro.net/GDB/show_report.asp?id=4614 by two all-terrain vehicle (ATV) operators riding on the western slope of the Blanca Peaks area in August 2000:

We didn't think too much of it until it moved. It stood upright and walked like a man. At first, we thought it was a hiker but it was all the same color, from head to toe. It walked about 15 yards before it walked into the trees. My uncle and I both stopped to make sure we saw the same thing. But we drove down the road about 300 to 400 yards before we decided to go look for it. We walked into the trees about 200 yards and came to a small meadow. My uncle was looking the other way when at the other end of the meadow it ran through. I yelled, 'There it goes!' We took off after it, on foot. This time I got a little bit better chance to look at it. The creature was a light to a medium brown and had shaggy long hair; it stood about seven feet tall. When we reached the end of the meadow, each of us went in an opposite direction. My uncle went the same direction as the bigfoot, and I went the other way in case it double-backed on us. But we didn't see the creature after that. We did find a few footprints, but didn't have any plaster to make a mold. So we went back down the mountain.

There have been over a dozen quality Bigfoot reports filed in and around the SLV including a flurry of reports investigated by Costilla County officials in 1994 and 1995, as detailed in Secrets of the Mysterious Valley, pages 229-232.

Over the years several intriguing reports have been filed relating to other strange cryptozoological animals spotted in and around Winchell Lake, just over the ridge from the Como Lakes. In 1997 I interviewed a prospector who claimed he had been camped at the lake in the eighties when he noticed something strange:

I used to spend a lot of time up in that area and I'd fish in [Winchell] lake. I'll tell you, there are some pretty weird fish in there. Funny thing, they're all deformed. I'd catch these big ol' thick fish that were only a few inches long, or these real long skinny ones. They'd have these real deformed heads and the strangest looking fins. Never actually ate one though, they were just too strange looking… There's some pretty strange things going on up there. I remember one night I was up at the lake, about an hour before dark, and I happened to look across the lake; it's not very big, not even a quarter-mile across, and what do I see? A huge white buffalo just standing there, plain as day! I wondered to myself, what in the world is a white buffalo doing all the way up here? Then it just disappeared! I went over there and couldn't find any sign of it, no tracks or nothing…

You know, I'm half Cherokee and half German, and I grew up on the Navaho Reservation as a kid. I'd hear all kinds of stories about this Valley. This is the extreme northeast corner of Navaho land, and to all the Indians who visit, this is a very sacred place. The Navaho believe they came into this world from that lake up on Blanca… I can only tell you what they told me, and they said we emerged out of that lake, high up on Blanca… I remember being told that the water broke through the top of the mountain and flowed down the southwest side. The Navaho people were told to live where the water flowed. They have lived there, on the western edge, for the longest of times.

Over the years other strange white-colored animals have been reported at the shoreline of Winchell Lake including an impossibly-large white buck deer. Other strange animals have been reported back over the ridge at Como Lake. In 1966 two local SLV fishermen, intimately familiar with the Blanca Peaks area, observed what they described as a "platypus" swimming slowly along the shoreline of the lake. The creature (normally only found in Australia) had the distinctive broad, flattened hairless bill and flattened tail. Startled, the two men matter-of-factly reported their inexplicable sighting to the American Museum of Natural History, but claimed their report was laughed at by museum officials and dismissed out of hand.

Another anomalous "creature" has been reported dozens of times in the San Luis Valley. Two-foot long, undulating apparitions— traveling close to the ground— have been seen repeatedly by witnesses in two specific spots in eastern Saguache County. These "prairie dragons," as one Navajo source called them, are shaped like flattened slugs without any apparent head; they are translucent at the ends but opaque in the middle, have no legs and leave no tracks or other evidence to indicate their presence. They have been reported coming out of walls and disappearing into walls and, curiously, these apparitions seem to be witnessed during time periods when a wave of UFO sightings and other strange anomalies is underway. Dogs respond to their presence and, at one location, they have been reported in groups of up to "a herd of 60" crossing the road. This 100-yard stretch of Road T, is located at the first set of turns heading west out of Crestone.

During Moffat High prom night 1993, several parents on the road allegedly observed a phantom herd of horses that galloped alongside their cars before veering off into the chico brush. As the startled drivers sped up, the three or four horses (depending on the version) kept pace with the vehicles without changing their stride. In this undeniably peculiar location on Road T, one of the witnesses reported seeing a strangely primitive, open landscape beyond the horses, although fence lines and two houses should be plainly visible from the road.

UFOS DEJA VU

Another type of phantom creature has been reported sporadically over the years in the SLV, and these unidentified flying object reports may help redefine our thinking about UFOs. The legendary "thunderbird," a huge, jet black, bird-like apparition has occasionally been witnessed around the remote Ute Mountain/Costilla, New Mexico area, on the eastern side of the Valley near the Colorado-New Mexico border. This majestic, soaring "hole-in-the-sky" monstrosity was most recently reported in the spring of 1993 by a ranch foreman tending his herd near Questa, New Mexico. It was also reported by a Mesita, Colorado resident and two visitors during the fall of 1986.

The Highest Incidence of UFO Sightings in America?

Reported UFO sightings in the San Luis Valley number in the thousands but there are a number of specific, documented reports that deserve mention.

According to a San Francisco newspaper article that appeared in September 1948, a San Luis Valley resident named Grant Edwards, Sr. had been showing amazing daylight UFO footage of multiple objects to civic groups around the SLV. It seems that in August of that year, he, his wife and two kids went out for a picnic on the banks of the Conejos River just north of the Colorado-New Mexico border. Edwards had just been given a new eight-millimeter movie camera and he couldn't wait to test-drive his new baby; it was loaded up and ready to go. That August afternoon, he unwittingly became the first US civilian to film multiple daylight UFOs. Edwards evidently presented the film to several dozen people before the film was appropriated by "the FBI" six months later in early 1949. Fifty years later, a neighbor, Marianne Brown (who witnessed the film multiple times as a teenager) remembered her feelings while viewing the footage for the first time:

I couldn't believe it. The UFOs would come down right over the trees and hover, then go off real fast, then the others would come down. They even flew in formation, with two together in front and the other three behind. At one point they hovered and turned on end like frisbees. The film was as clear as could be! [Edwards] did a good job of filming them. When they hovered low over the trees, you could even see his wife and daughter standing in the foreground watching them.

I received confirmation of this potentially historic film from Grant Edwards' son who, at the time I interviewed him, was a County Commissioner. In a reluctant manner he corroborated the film's authenticity, telling me on-the-record, "Yes, my Dad was the same Grant Edwards who filmed the UFOs."

From 1992 through 2000 almost every conceivable type of UFO craft was reported numerous times in the San Luis Valley. Red, orange, green, white and blue "orbs" were seen along with all sorts of traditional saucer-shaped craft. Huge black triangles were reported in one incident, accompanied by "military helicopters."

UFOS DEJA VU

Photo by Michael MacLaughlin >

Zapata Falls
August 17, 2007

Space constraints prohibit even a quick overview of the hundreds of documented UFO sightings in the SLV, but research has uncovered what is alleged to be one of the very first documented UFO reports in North America worthy of mention.

According to archaeo-astronomer Marilyn Childs, the diary of New Mexico Territorial Governor Juan Baptiste de Anza contains a 1777 entry that describes strange lights flying around Blanca Peak, along with a description of a powerful low humming sound heard emanating from the mountains. At the time, De Anza was leading an army that was chasing renegade Comanche Chief Cuerno Verde through the SLV. I have not been able to corroborate her assertion.

Other reports from the late 19th century and early 20th century suggest undefined aerial phenomena have been witnessed in the SLV and elsewhere for generations. These reports, made prior to the first conventional aircraft flights in the regions, obviously cannot be chalked up to "misidentified military activity." Anecdotal research of these isolated historical events cannot prove or disprove their high-strange nature, however it appears that a long time ago something intelligent, with the aid of high technology, seems to have singled out humanity for unknown purposes. In today's troubled skies, the plot has thickened; for now it appears we have new human players joining the game with unstated agendas.

The NORAD Events

As noted in a recent article here at ourstrangeplanet.com, the NORAD Events (possibly the USA's "dark horse" UFO affair of the 1990s) was comprised of a series of

160

UFOS DEJA VU

unexplained occurrences over a six week period, including reports of various types of UFO craft; two green, two blue, two orange and one white fireball; two orange orbs; two huge, mystery high altitude fires; a variety of mysterious booms; a flurry of Bigfoot reports; a documented unusual cattle death; and dozens of reports of accompanying militaryesque activity. This flap began in early December 1993 and continued until the early evening of January 17, 1994.

The most sensational series of events occurred during the afternoon of January 12, when a NORAD official contacted the Rio Grande Sheriff's office at 3:40 P. M., and reported "a significant explosion" logged at 2:55 P.M. in the Greenie Mountain-Rock Creek Canyon area by a NORAD satellite scope operator in Cheyenne Mountain.

Exactly two hours later to the minute, at 4:55 P. M., Florence, Colorado resident Lt. Col. Jimmy Lloyd (ret.), a 30-year veteran fighter pilot and self-professed UFO skeptic, reported seeing "a battleship-sized ... glowing green" group of "six or seven objects in close (crescent) formation" streak overhead just south of him. They then appeared to "go down into the San Luis Valley." According to Lloyd, the objects were not mundane celestial objects, e.g., meteors, missiles or any type of conventional craft, and were completely silent.

This incident shares several aspects with the 1983 Gallup Incident that occurred exactly 11 years earlier. There, too, explosions were heard first and fireballs were seen later with the same two-hour time lag! For details of that incident, see Secrets of the Mysterious Valley, pages 239-242.

Aerial craft or helicopter sightings and rumors of military ground activity were

161

reported around the Greenie Mountain event, combined with probable misdirection by the government. Maj. McCouch, FEMA supervisor of the NORAD scope operator, appears to have directed local search crews to investigate a rugged area nearly 25 miles from the probable impact site, as indicated by the original coordinates. Was this suspected misdirection given to allow for a military search of Greenie Ridge with the low-flying B-52s and helicopters people reported seeing during the following four days? Two UFO Institute investigators in the area reported finding heavy equipment and snowshoe tracks; they also claim to have stumbled on huge "metal doors" in the ground a week after the NORAD story broke. The Greenie Mountain area is dotted with closed up, abandoned mine shafts, and rumors of antiballistic missile sites abound. To my knowledge, there is no verifiable evidence of missile site activity or retrieval operations during the NORAD Events. Rumors aside, it would appear some other agenda was at work.

When the crew of Varied Directions, a TBS film production company, called me a couple of months later they told me that they were planning a trip to the San Luis Valley, and then on to NORAD in Cheyenne Mountain, near Colorado Springs, to investigate the strange call to SLV law enforcement. The military initially granted permission for them to visit the secret mountain base, but when it was revealed that they were interested in the SLV events, their visit was promptly canceled for no apparent reason. It is noteworthy that Varied Directions was the first civilian film crew allowed aboard a Trident submarine when they produced a documentary on the sub for the PBS program Nova. They had also worked closely with NASA on a documented history of the space race called Moon Shot—a feature-length documentary based on the book written by ex-astronaut/Mission Control boss Deke Slayton. This is obviously a well-connected production company. But even with these impressive credentials and an inside track with the military, permission was revoked to film inside Cheyenne Mountain by NORAD officials.

The full story of the military's involvement and the extent of their presence in the SLV will never be fully known, but it is safe to say that the government is extremely active in the region. It is also probable that many so-called "UFOs" sighted in the area are really conventional and non-conventional military craft. The obvious correlation between military ground activity and waves of unusual aerial object sightings is highly suspect in this investigator's opinion.

The Unusual Livestock Death Phenomenon

In the realm of the paranormal, the San Luis Valley is most notorious as the "birthplace of the cattle mutilation phenomenon." The first widely publicized case of this type occurred right in the heart of the SLV, but what makes this distinction compelling and perplexing is that the case in question featured a horse. In the thousands of cases of animal mutilation reported since, no other remains have ever been reported in the same horrific condition as those of "Snippy the Horse."

UFOS DEJA VU

Not a popular subject with the casual paranormal crowd, ever since the Snippy case occurred in September 1967, the scourge of cattle mutilations has quietly spread around the Western beef-eating world. Since 2002, over 3,000 cases have allegedly been reported in Argentina alone, and estimates of the pervasiveness of the mystery most often cite the number of cases worldwide to well-exceed 10,000. Since 1967, around 200 official reports have been filed by angry, puzzled SLV ranchers, but the total number of cases may be closer to 1,000. Although "misidentified scavenger action" may explain a few of these reports, most ranchers are skilled, knowledgeable outdoorsmen who know what is a mundane livestock death and what is truly high strange. It stands to reason that they would not invite the scorn often associated with claims of animal mutilation upon themselves and their families. Couple this with dozens of reports of unusual military-style helicopter activity in and around mutilation sites and you have a truly puzzling scenario that is not easily studied, debunked or denied.

Stalking the Herd

For more in-depth information on the enduring "mutilation" mystery, my recent book, Stalking the Herd, examines our close relationship with bovines that extends back at least 35,000 years and how this collective relationship may be manifesting the modern "cattle mutilation," phenomenon. And what about the possible covert monitoring of the food chain for prion disease ('mad-cow)?

UFOS DEJA VU

World class investigator Chris O'Brien.

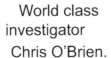

Center: Michael Mclaughlin is responsible for taking this photo of a morphing object.

< What is this in the sky? Photo by Tim Edwards of a mysterious "cigar" over the Valley in 2007. UFOs have been known to change shape as this picture proves.

UFOS DEJA VU

THE UFO DATA ACQUISITION PROJECT
By Timothy Green Beckley

PUBLISHER'S NOTE: With the ongoing widespread UFO activity in the Valley, Chris O'Brien has long recognized the need for constant aerial surveillance to record and preserve the essence of such reports.

"In early 1993 when I began what became an almost 10 year intensive investigation of unexplained events in the San Luis Valley, Colorado, one of the first things I imagined was setting up a surveillance net of cameras around the SLV to attempt to capture on tape (we still used videotape back then) the many sighting reports that, at times, poured fast and furious into my tiny mountain community of the Crestone/Baca Grande. Several of my early sightings were so spectacular that I was seriously inspired to cost out what a high-tech system would entail. It didn't take long to realize that this vision was impractical for several reasons.

"First, static camera placement automatically meant you'd either needed to be psychic, or you had to be extremely lucky to have cameras in the right place at the right time. Second, how would they be able to go from an idle position into full blown record mode? Without that capability, you would be bogged down with 4-5 tapes per day that you would have to review, and no matter how much activity we were having at the time, this was extremely impractical and cumbersome. Plus, having the cameras in the right place meant a certain amount of needed portability and, at the time compounding the problem, good quality video cameras were rather large and bulky. Then there was the question of weatherproofing them; the list of problems and prohibitive cost seemed insurmountable and as I told several researcher friends at the time, "I'll bet this kind of system would be feasible in the not-to-distant future. Meanwhile, from 1992 to 2002, I logged hundreds of UAP/AAO/UFO sightings in the greater San Luis Valley."

But Chris was not about to give up when it came to setting up some sort of system that could record any activity of potential interest and importance

"Fast forward 13 years. Welcome to the future! In 2007 I met ex-aerospace intelligence agency contract lawyer Wayne Hollenbeck at a special invitation event I arranged for Ray Stanford. Wayne, who knew about Sedona's reputation for fairly consistent UFO sighting activity had brought a high-tech Sony surveillance camera with the idea of convincing me to set it up in my area to attempt to capture some sightings and transmit them to him in California via the Internet. It was then that it dawned on me that in 2007 with the Internet my vision of surveilling the SLV had a chance off being realized. By 2010 we had a camera atop a 100 foot tower in downtown Alamosa, Colorado—the largest SLV town. Unfortunately, the building was sold after the camera had only been up a couple of months, and we had to take it down, but we did manage to capture several interesting clips.

"One clip shows a small two person helicopter entering the frame midway up the right side of the frame. If you notice, the sun hits the candy at just the right angle, and the chopper blooms into a perfect silver sphere and sails across the middle of the screen from right to left. Now, if I was unscrupulous and had compromised intentions, I could have simply cut out the first few frames and voila', I would have had sensational silver UFO sphere footage. But, as fans of my work and my friends have known for a long time, that's not why I'm involved in the serious research, I want answers and I also want to help educate the public as best I can."

THE DREAM COMES TRUE

Eventually Chris O'Brien's dream of 24/7 aerial surveillance became a reality. Fast forward to September 2018. After several false starts, the SLV Camera Project has joined forces and combined efforts with software/computer engineer, inventor Ron Olch. Ron, a longtime computer scientist with Disney is our newest team member and the project has now morphed and fine-scoped as a potential worldwide effort now called UFO DAP which stands for Data Acquisition Project. What is UFODAP? Well, I can safely say that our project is a major leap forward for the sorry field of "scientific" ufology. The study of UFOs is in woeful shape. If ufology was a car, we have been trying to drive it forward by using the rearview mirror to try and see where we're going. For far too long, the field has been fixated with old, anecdotal cases and questionably viable database mining. Garbage in/garbage out, as they say. It's about time that a group has devoted the necessary time, talent, and money to come up with a real potential for real-time, hard data acquisition of scientific principles behind UFO events. Welcome to the future!

AN OVERVIEW OF THE UFO DATA ACQUISITION PROJECT

"For 70-plus years, the so-called field of "scientific ufology" has been a misnomer for truly viable UAP/AAO/UFO research and investigation. The effort to ascertain the nature and capabilities of these mysterious objects has been stymied by the mercurial aspect of their tricksterish manifestation, and the prohibitive cost of viable, diagnostic technology. What we've been left with, for decades, is a bewildering, growing pile of anecdotal reports that are impossible to analyze effectively. Something has to change, and perhaps this perplexing conundrum may have finally been solved.

'UFO Data Acquisition Project (UFODAP) in the San Luis Valley, Colorado involves the deployment of pan/tilt/zoom video cameras and multi-sensor data acquisition sensors to properly record real-time UAP/AAO/UFO events. These instruments will document anomalous aerial objects at our first location, America's most active UFO "hot spot," the San Luis Valley, Colorado. The system will grow from two initial cameras to a three camera system with data sensors to be added as time and funding permit. Multiple cameras and sensors allows for all-important "triangulation." The triangulation feature will permit an evaluation and determination of object size, distance, altitude, speed and acceleration."

A report on the UFODAP web site indicates that they have already generated great interest in the anomalous aerial object investigation community. Inquiries from Switzerland, Australia, Italy, Canada and across the United States have already been made by on-site researchers located in "hot spot" areas. Here is a quick list of several potential sites the UFODAP team would like to see monitored with 24/7 video cameras and Multi-sensor data acquisition units (MSDAUs)

** Brown Mountain —

** Oscola Peak, NM -

** Bridgewater Triangle (Massachusetts) -

** Estrella Mountains (Arizona, Phoenix Lights) -

** Mount Adams Triangle (Yakama, Washington) -

** Chestnut Ridge (Pennsylvania) —

* * * * *

A full report from the www. UFODAP.com site is reproduced below.

It is apparent that timely collection of high-quality optical and electromagnetic scientific data related to UFO events has been difficult to obtain. Individuals who are in the position to potentially record such events often do not have the appropriate equipment at hand. Even MUFON Field Investigators may not have the means to wait for hours, days or longer to capture an event and, at that time, record all of the necessary data in a verifiable way.

Even when photos or videos are recorded, they often lack verifiable associated metadata such as the exact location of the camera and sensors, the azimuth and elevation of the where the camera is pointing, time of day, associated electromagnetic and gravitational perturbations at that time and so on. Also, even if a single camera captured such data, the track of an object, its location on or above the Earth and its altitude, could not be ascertained without combining the data of at least two such systems, placed some distance from each other.

The focus of the UFODAS development has been on resolving a portion of this issue by providing methods to recognize, track and photograph anomalous ob-

UFOS DEJA VU

jects while simultaneously collecting data from multiple sensors. While this sort of capability has been investigated and other systems have been built, their design emphasis has not been on such low cost to make practical the kind of significant numbers to be deployed to have a practical impact on Ufology. By "low-cost" we assume a unit cost of perhaps $2000 or less. Thus, over the last five years significant progress has been made on an Unidentified Flying Object Data Acquisition System (UFODAS) that attempts to address this issue.

UFODAS consists of a Windows operating system-based personal computer and options of one or two cameras and other clusters of sensors. In addition, there is software to pull data and video, locally or over the internet, from multiple sensor locations and triangulate target objects. The system supports a wide-range of supported cameras including USB webcams up to sophisticated all-weather IP cameras with pan and tilt as well as optical zoom. The software architecture is designed to adapt to most any camera or Pan-Tilt-Zoom (PTZ) mechanism in the future by addition of a single software element, without modification to the main UFODAS software. In dual camera applications, one camera may be a non-PTZ type that views a wide field of interest including all-sky cameras. The second camera would be a PTZ camera directed to point at the object based upon its relative location in the field of view of the wide angle camera. The PTZ camera then independently tracks the object. Whether using one camera or two, the processor samples frames from the wide-field camera and through some fairly sophisticated image analysis, detects qualified moving objects. It then directs the pan-tilt head to point the telephoto camera at the object and collects images from it. The software is capable of acquiring an object of interest and smoothly tracking and zooming a moving object even with a single camera. Maintaining track while moving the camera, which causes the background to also move, was a significant part of the development effort.

The software architecture employed enables support for additional cameras, whether simple or sophisticated, including those with fast PTZ operation, higher resolution or non-visible spectrum devices.

Triangulation of a sighted target object requires accurate azimuth and elevation of the tracking camera. The optional MultiSensorUnit (MSDAU) is an embedded hardware and software subsystem that provides camera GPS coordinates and precise time as well as 3DOF magnetometer and DC accelerometer. The same sensors may be used to sense perturbations in those fields and include that data with a camera-based event or actually provide the initial trigger for subsequent data collection.

The software also provides a number of related functions which include:

· When a qualified event is detected, sends an email to a designated address with data that includes attached photos, GPS coordinates of the camera and object azimuth and elevation.

· Upload collected data to a Google Drive or Dropbox account.

· Saves automatically named photos and videos to folders it creates in local memory.

· Operational parameters can be adjusted by sending an email to an UFODAS-specific email address.

· A sophisticated Graphical User Interface (GUI) for user-friendly operation.

Ongoing development work includes:

· Additional methods to eliminate false alarms such as birds and aircraft including the use of deep learning methods.

· Real-time RF spectrum analysis option for the MSDAU with "video" capture of the changing spectrum recorded by an MC.

· MSDAU interface for acoustic sensors.

· MSDAU interface for radar data.

· Real-time track correlation with data from the flightrader24.com website to distinguish unknowns from aircraft.

· Use of an optical gradient filter to determine target spectrum.

· Differential magnetometry to determine target magnetic field strength and direction.

The UFODAS system architecture provides for an extremely broad set of configuration options to meet the goal of providing systems for every budget and type of case.

UFODAS architectural components consist of:

· Mission Control (MC) GUI-controlled software. MC interfaces with other elements via the Internet to bring together, in one location, data from up to six Data Acquisition Units (DAUs). DAUs may be any combination of OTDAUs or MSDAUs.

· Optical Tracking Data Acquisition Unit (OTDAU). An OTDAU includes a GUI-controlled software element that provides an interface to many types of cameras for optical target acquisition, tracking and video storage. An OTDAU can either stand-alone or work with MC. Two OTDAUs and an MC form a comprehensive solution to tracking with triangulation and both OTDAU and MC local data storage.

· MultiSensor Data Acquisition Unit (MSDAU). An MSDAU consists of an all-environment enclosure with an embedded Raspberry Pi computer interfaced to nine different sensors including GPS, magnetometer, DC accelerometer, AC accelerometer, temperature and pressure. An MSDAU communicates with an MC over the Internet to provide all of this data in real-time. An MSDAU may also transmit data from other USB-interfaced sensors such as a Trifield meter.

· A combination of a PTZ camera mounted on top of a MSDAU which is then

UFOS DEJA VU

tripod or wall/pole mounted. In this configuration, data from the camera and sensors is combined into a single Ethernet cable for communication with MC. The MC then can be configured to use the co-located MSDAU data to locate the camera and collect multi-sensor data simultaneous with tracking events.

UFODAS cameras, the MSDAU, OTDAU and MC software as well as numerous installation and support options are available via the UFPDAP website. Four cameras are offered, each with unique capabilities and price levels ranging from a fixed lens, wide-angle unit, an All-Sky 360 degree camera to 12x and 30x PTZ models. OTDAU software allows the use of each type alone or in combination.

The sand dunes in the San Luis National Park seem to draw UFOs like a magnet.

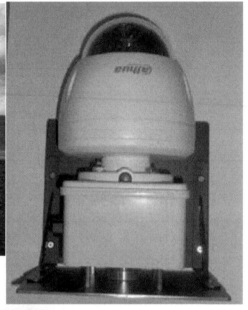

Above: An entry level data system can be purchased for less than a thousand dollars and will enable researchers to record UFOs.

Left: For around a thousand dollars a DAP system can be installed at any UFO hotspot.

It is hoped in the next few years that such equipment can be installed at a variety of locations assumed to be vortexes.

UFOS DEJA VU

FIREBALLS, BLACK CHOPPERS, FACELESS PHANTOMS
AND LEGENDS OF WITCH MOUNTAIN
By Brent Raynes

PUBLISHER'S NOTE: Seems our spy in the sky, Brent Raynes, always efficient, learned about the bizarre nature of the San Luis Valley and surrounding areas early on and set out to discover for himself the what and why of the unexplained phenomena transpiring on much of what can be considered to be "sacred ground."

He learned, as you shall, that there are fireballs that materialize seemingly out of "nowhere," helicopters that fly at all hours of the day and night, faceless specters that stalk the byways, as well as the legends and lore of haunted mountains. As a plus, Brent invites us to load up our cosmic vans and join him on the road – next stop, a portal just around the next bend.

* * * * *

Today as we facilitate our study of the paranormal, we have found ourselves confronted by the awkward merging of old and new beliefs, and a combining of many aspects once thought separate. Such a trend offers a considerable challenge to a modern intellect's cognitive sensibilities, certainly surfaced very noticeably during our recent sojourn out West. Talk of UFOs was interspersed freely with tales of Bigfoot, skinwalkers, demon dogs, shapeshifting witches, and encounters with the devil himself. Lurking in the shadows of all this high strangeness are ominous black ops types, or at least that is what some folks are saying. Some of those folks are admittedly a little "out there," if you catch my drift, while others seem ordinary, down-to-earth, and decidedly credible. One Colorado resident described to me how she felt certain that two men in a black unmarked helicopter had attempted to abduct her right in the city limits of her hometown of Alamosa. Nearby Mount Blanca, regarded as a sacred place of emergence by the Hopi and Navaho, is suspected by quite a number of UFO believers as harboring a concealed military base of operations that monitors the local UFO activity. UFO activity in this area is said to be quite high. There certainly have been a lot of strange and puzzling reports in the San Luis Valley of Colorado over the years.

UFOS DEJA VU

I had never been to New Mexico or Colorado, so I was certainly very excited about this trip, as was my wife Joan. Flying into Albuquerque, New Mexico late Thursday night, August 7th, 2008, we were met by Priscilla Wolf and her dear friend and long time companion Steve. They greeted us as we were headed to pick up our baggage. The way we conversed and laughed a stranger watching us doubtless would have assumed that we had all known each other and been close friends for years. But the truth of the matter is that we began communication with one another via the Internet, snail mail and telephone back in April of this year. I owe a solemn debt of gratitude to that indefatigable researcher and prolific writer, Brad Steiger, for ultimately bringing us together.

Though the trip to their home in the mountains of Tijeras, east of Albuquerque, is approximately twenty miles, it seemed like no time at all till we arrived at their beautiful mountain home/retreat. However, not long after arriving the adrenaline was beginning to wear thin from the long trip, which our gracious and perceptive hosts quickly picked up on, as they, introduced us to our sleeping quarters. They told us to feel at home and they certainly did indeed make us feel that way.

POWERFUL EARTH ENERGIES

The next morning we proceeded at a relaxed pace. A healthy and very delicious home prepared breakfast was set before us as we found an endless variety of subjects on which to converse. Day one was going to involve a trip over to Albuquerque, to the west bank of the Rio Grande where the Petroglyph National Monument is located. I had proposed this particular field trip to Priscilla a few weeks earlier and had even emailed her a copy of an interview I had done last year with New York researcher and writer John Burke, author of "Seed of Knowledge, Stone of Plenty." Priscilla was very impressed with Burke's work and even began exchanging emails with him. Burke had visited ancient mounds, henges, pyramids, stone chambers, and known sacred spots at such diverse places as England's Silbury Hill and Avebury Henge, Guatemala's oldest Mayan city of Tikal, the Black Hills of South Dakota, to name but a few, and using a fluxgate magnetometer, a standard voltmeter, and an electrostatic voltmeter, he and his colleague Kaj Halberg repeatedly detected unusual earth energies at these places.

"Overwhelmingly, the ancient megalithic architects all over the world chose to build on conductivity discontinuities, and then designed and built these enormous structures in such a way as to further concentrate the natural electromagnetic energies present at these sites," Burke explained in our interview. He further expanded, "A conductivity discontinuity is simply the intersection of two zones of land, one of which conducts natural electrical ground current relatively well and the other less well. At such sites the normal daily fluctuations of the earth's geomagnetic field are magnified several hundred percent, and with them the telluric currents that flow through the ground." Burke also pointed out that Canadian neuroscientist, Dr. Michael Persinger had "confirmed that the magnitude of magnetic changes we have found at these sites conforms to those he has found capable of creating visions in volunteers in his lab."

At any rate, the largest conductivity discontinuity that Burke had ever studied was (you guessed it) the Petroglyph National Monument! As he explained in our interview, "it contains thousands of rock carvings which are considered by anthropologists today to have been made by shamans illustrating their trance hallucinations. I measured very powerful and extremely odd surges of electric current in the ground there. When the ranger at the Visitor Center heard what I was finding, she said to me, "You know, periodically I get these 'New Age types' coming in here and telling me they just love to go sit up in the rocks and feel the energy. I thought they were a bunch of flakes, but you're telling me there might be something to this."

These thousands of ancient petroglyphs are scattered across the face of a 17-mile long West Mesa escarpment that was the result of volcanic eruptions that began about 150,000 years ago. There are said to be more than twenty thousand images covering many of the countless volcanic rocks at this immense site, images said to have been created four to seven centuries ago. Most of these petroglyphs are found on south and east facing slopes. We visited the Boca Negra Canyon portion of this site, walking what is known as the Mesa Point Trail. Boca Negra Canyon was the area where Burke took most of his measurements.

While we were walking slowly up the volcanic rock-strewn hillside we were startled when Priscilla's plastic water bottle made a pop sound and she cried out that she had just gotten an electric shock. "When I was going to put the top back on the bottle it shocked me and I put the lid on and it made a loud pop, (as) you heard," Priscilla recalled later.

We came upon one large black volcanic rock on the trail that had a place in it just perfect for putting our heads inside. "It fit all of our heads," Priscilla recalled. "That was the most amazing thing." She placed her head inside, trying to get rid of a headache. With her eyes closed and her head inside the rock, Priscilla said that she could see various colors, but mainly purple. I remembered myself "seeing" a green glowing cross shape, it's four appendages appearing to be of equal distance. Later, on another part of the slope, I pulled three whistling vessels from my bag. These are fully functional replicas of the ancient Peruvian Chimu vessels, believed by a number of researchers to have been shamanic instruments of sound. I was going to use them in my presentation that upcoming weekend (August 9th and 10th) at the UFO Watchtower Conference in Hooper, Colorado, where both Priscilla and I were scheduled to speak.

They are great for inducing profound altered states of consciousness, and I was interested in observing what impressions might be produced at a site as obviously powerful as this one. After a few minutes of blowing the whistles, Priscilla had a report for me. "I seen the energy of lights, first like flashing lights, and then it was like a sun rotating itself, coming down, like moving towards me," Priscilla recalled later. "I could see it just rotating around and around." I reminded her of what she had told me at the time about her back injury, to which she replied: "Oh yes. I just hurt my back a few days ago and I've been in a lot of pain and I thought, 'Oh no, we're going to be heading for this UFO place and my back is hurting so bad,' and I thought that I had probably damaged my back by lifting up the microwave. I wasn't too sure what had caused it, and on the vision today, during the whistles, I saw that I was real mad and that I had allowed an opening to come into my body, because I had been wanting that microwave moved. What happened is the anger had allowed an opening into my body to cause pain, so that's when I seen light and I used the light of the sun to close that."

EARTH SPIRITS AND ENERGIES OF THE SAN LUIS VALLEY
AND THE GATES TO OTHER WORLDS

A couple days later, late on the afternoon of Sunday, August 10th, as we were leaving the UFO gathering in Hooper, I would wonder again about unique earth energies and their possible connection to volcano sites. At our departure from there, Priscilla, in our drive into the southern San Luis Valley of Colorado, wanted to show us her childhood home, and where many people had experienced many strange things down through the years. As we rode down this dusty, gravel road of hills and prairie land, an extinct volcano drew closer and closer. Conspicuously absent was the presence of homes and other people, as we drove on mile after mile through this region. "That's a portal, what they call an opening, of 'Gates to Other Worlds,' right here," Priscilla said. "Right in this field?" I asked. "Right in there, and the markers are one, two and three hills right there," she said, pointing out three nearby hills located around the volcano.

"And you used to come up here as a child with your brother and you used to see the disks?" I asked.

"Yeah, the black disks flying around and just totally going into the volcano, disappearing in there."

"When they disappeared into the volcano, did they go down into the top of it or did they go into the side?" I asked.

"No," Priscilla said. "I don't know where they went to but they disappeared right into it. Like when you're going straight into the mountain. Never went up or down." Priscilla recalled climbing to the top of it and looking down into a crater. "Just rock like we'd seen over there at the petroglyphs," she added. I couldn't help but wonder what sort of readings Burke's instruments would pick up at this location.

"This is all Cerro De Las Brujas. Witch Mountain," Priscilla said at one point.

UFOS DEJA VU

"Witch mountain?" I said. "Yeah," she replied. "The Catholics and all the Mexicans named it that because of the weird stuff here. Nobody will live up here. No ranchers. Nobody. A lot of land." She pointed in the distance, "There's a camp over there of sheep herders." Then she muttered aloud, "Thirty miles of nothing." As we continued on, we saw exactly what she was talking about.

Priscilla shared so many strange stories about this region, that it's hard to know where to begin. Her family knew a Ute Indian who had for years been a sheep herder in the Cerro De Las Brujas region. This allegedly happened back in the early 1900s. "He had a lot of experience with the black UFOs," Priscilla explained. She recalled one instance where he described seeing a "dust devil" that emerged from "this weird craft" and this "dust devil" looking thing "took him up" and then "the next thing he knew he was in these people's home. ...it really was like back East. They were having a big party and there were a lot of white people. There was a lot of green grass. The place looked different than what we [had] ... When he moved outside they offered him food. He ate the food and everything. It was something that he thought was the future. Then, all of a sudden, he looked back and there was like a flash and he was back on the prairie, back here. He had been missing for several days and they had been looking for him, but to him he had just been there and gone and come right back."

Priscilla recalled another incident that he had described. "In the part where the volcano mountain is at, that's where they would appear at," she explained. "Another time, there were three flat white disks and he spotted them out there and they disappeared into the volcano mountain, like they went into another dimension, and then the man in black was walking around."

"Around the base? [of the mountain]" I asked.

"Around the whole area that we went to," she said. "He was just walking around after the three white disks went into the mountain."

"Like right inside the mountain like?" I said.

"Yeah," Priscilla stated. "Where they call it Gates to Other Worlds."

"Then the other time it happened to him and it was like he was taken into the past and there was a huge house," she recalled. "This huge house is not there any more. Close to Witch Mountain. There were a lot of people in a circle praying to the devil, doing devil worship, and this man who was half human and half goat, he said he was the devil, started dancing in the middle of the circle. Then he bent down and asked all of them to kiss his ass. When he got to him he asked God to help him, and he disappeared and all of a sudden he was just back, all dusty and dirty, with the dirt there back around his camp."

Priscilla's grandfather, also a sheep herder, recalled going down to the river to collect some fire wood one night. "This was the early 1900s," she said. "Right there across the river, on the Rio Grande, a big black dog appeared and wouldn't

175

let him cross and the horses were backing up. He took his rifle out, and he always kept it loaded with silver bullets, and he made a cross on it and shot it and when he shot the black dog it turned into an old woman. So he said it was a brujas, a witch turned into a black dog." Priscilla recalled that the Ute Indian sheep herder "had also seen the black dog there after sightings of these weird flying objects."

I asked Priscilla to describe this mysterious "dog" a little more. "It's not a normal size," she said. "It's a huge black dog. Fire comes out of his eyes. Real red eyes. Then he vanishes. It's never been known that he hurts anybody. It's just that he appears and scares people."

A FIREBALL AND A FACELESS HITCHHIKER

Another curious anomaly described in the area are "fireballs" that come "rolling down" from the tops of nearby hills. "They look like they're coming at you and they're going to hit you, and then right before they make it to you they disappear," Priscilla said. "I had a lot of experiences with those energy balls, I call them. The ones that I saw were like red balls of fire. Some of them were white, yellowish – like electricity." In a few instances they apparently struck motor vehicles, but then they'd "just vanish." No sound and no heat, and, Priscilla added, "There was never an explosion."

Then there is the phantom faceless hitchhiker! "I know some ranchers here have seen it," she said, "I know members of my family have seen it. A lot of people have said that it's an old man in raggedy clothes barely walking and they stop to ask him [if he needs] a ride or they'd even honk at him to get off of the road, and then when he turns around his whole face is skeleton."

We also stopped and took pictures of a mound of huge boulders next to the road. People claimed that animal sacrifices to the devil had been performed there. Some even alleged human sacrifices had also been done.

Recently Lightning Bolt, a Native American spiritual teacher, counselor and healer from the West Coast has engaged me in an ongoing dialogue about the UFO/shamanic connection. "You are right, there is a connection between 'shamans' and UFOs," he wrote. He explained that he's had many personal UFO encounters. Though we hadn't been specifically discussing volcanoes, he wrote: "Have you ever wondered why so many UFO sightings happen around volcanoes? Mt. Baker, Mt. Saint Helens, all the way up into Canada/Alaska down into the volcanoes in Mexico. They gather energy, including radiation, out of the volcanoes to power their ships. Volcanic eruptions emit a lot of radiation. Did you know that? The UFOs know how to gather the radiation for a source of power."

All of this dark and sinister history aside, when we first entered the San Luis Valley, on our way to the UFO conference, Priscilla gave us a little history lesson about the Native American history of the region, which illustrated the area's predominantly positive characteristics. She explained, "The Ute Indians, the

UFOS DEJA VU

Comanches, the Navahos, Apaches, the Pueblo Indians, years before the white man came, would go here to hunt and fish. They had plenty of elk, deer, bear and fish. Nobody fought here. It was called The Bloodless Valley. The Great Creator would not allow any blood shed. It was a place of energy. They could kill animals, but killing human beings was not allowed." The region is surrounded by a majestic mountain range. One of these mountains, Mount Blanca, is considered very sacred to the Indians. "That's where a lot of the Navahos and the Hopis claimed that they emerged from the ground, and they were taken care of by the ant people, the little people," Priscilla noted.

Before the conference, we stopped in Alamosa and I got to meet Priscilla's sister Angela. I had interviewed Priscilla about Angela's haunted home and Priscilla's skinwalker encounter there. Angela told me herself how she, too, had seen the entity, how she had come home from the hospital and then saw it crawling toward her, it's long black panther like body, but with a head like a man.

In New Mexico, Priscilla recently spoke with a real estate lady who "got real scared" when the subject of skinwalkers came up in conversation, and admitted, that she didn't like talking about them, but that her brother and someone else encountered one together while out riding. "The skinwalker was half man and half wolf," Priscilla recalled. "The faster they went he kept up with the car, looking at them, on the side of the car. It really scared them to death. They were going as fast as 50 and probably a higher speed than that. It's well known about skinwalkers appearing to people on Route 66. That's nothing new."

Incidentally, Priscilla lives just off Route 66.

Do the skinwalkers worry her? She says, "You've got to know how to really protect yourself. If you have the protection of a medicine bag, a silver cross, and you're religious, they can't touch you. They can't take your soul. I'm not going to say I never get scared. My God, it's a miracle I haven't had a heart attack and died right there. Because all of a sudden you see something that's half human and half animal."

A VOICE OUT OF THIN AIR

Friday evening, the day we had been to the petroglyph site, we had returned and were relaxing and sitting at the dinning room table. Steve was sitting at the other end, and Priscilla was to my left and Joan to my right. It was approximately 7:25 P.M. when I heard a male voice yell "Brent!" It seemed to be coming from outside, in through a screened window to my left. I looked at the others seated at the table close by, expecting them to comment or chuckle over this, but instead everyone gave me this puzzled look back as to why I was looking at them the way I was. I soon discovered that not one of them had heard what I had heard! I then asked Priscilla if it would be okay if I excused myself for a moment to go get my tape recorder and begin taping some of her stories. She was agreeable to doing that. Soon I returned, at which time Joan and Steve had gone into another room to talk among themselves, while I questioned Priscilla about her stories. I also wanted to be prepared in case I was to hear that voice again. I was a little puzzled as to why no one else said that they had heard it. I began asking Priscilla if this sort of thing happened to her. She admitted that she occasionally heard her name called. I asked if it came from any particular side. "I usually hear it from my left side," she explained. "....what's weird that's my hard of hearing side." Recently, as I was transcribing this tape I came to a part of the conversation, that same evening, where I thought that I had seen a cat run by the table. Priscilla seemed to think it was just one of the house cats, but then she added, "I do have a ghost cat that passes by. No really, honest to God," and then she laughs. Right after that, on the tape, you hear a female sounding voice, quite clearly say, "Brent!" But we keep talking like we didn't hear a thing, oblivious to this voice. So this time the tape recorder heard my name being called and I missed it!

I was 56 years old at that time, and I'd been to lots of haunted locations and spoken with a lot of psychics, and I'd never had (to the best of my recollection) this experience of someone calling my name whom I couldn't identify. Many people have told me about hearing someone call their name, how there's no one there or no source for it, but it never happened to me. Never, that is, until the evening of August 8th, when it happened twice....the second time as a possible EVP.

This fabulous book published by Inner Light contains numerous interviews by Brent Raynes with Native American shamans who discuss their unusual experiences with UFOs among other paranormal topics.

UFOS DEJA VU

A closeup of a Native American petroglyphs in New Mexico's part of the
San Luis Valley.

Joan Raynes touches Native American rock art at the
17-miles mesa in New Mexico.

UFOS DEJA VU

Pricilla Wolf has undergone many supernatural experiences in the San Luis Valley

BRENT RAYNES

If you're any sort of UFO buff or researcher you will want to get a complementary subscription to Raynes' online newsletter Alternate Perceptions www.apmagazine.info

UFOS DEJA VU

< On the left, Paul Bennewitz, an electronic engineer from New Mexico, who argued, around 1979, that there was an extraterrestrial base close to the town of Dulce (New Mexico) involved in cattle mutilations and abductions in the area.

On the right, one of the photos of the Bennewitz collection, where you can see a strange ship flying near the Dulce area, the Kirlkland Air Base and the Manzano Nuclear Weapons Center. Many of the UFOs Paul saw vanished right through solid rock. >

UFOs have been known to go right through the solid walls of mountains according to Pricilla Wolf.
Photo by Paul Benewitz in New Mexico.

UFOS DEJA VU

Rock climbers scale Witches' Tit, San Luis Valley.

SECTION FIVE MULTIPLE PHENOMENA

UFOS DEJA VU

CHAPTER TWENTY-TWO
MULTIPLE PHENOMENA ON A COLORADO RANCH
By Dr. John Derr and Dr. Leo Sprinkle

UFOS DEJA VU

MULTIPLE PHENOMENA ON A COLORADO RANCH
Report on the Investigation of UFO Experiences
on a Rocky Mountain Ranch
By Dr. John S. Derr, Ph. D., APRO Consultant in Seismology
and
Dr. Leo Sprinkle, Ph. D., APRO Consultant in Psychology

PUBLISHER'S NOTE: This is one of the most complete reports ever filed on an ongoing UFO flap at one particular locale. A perfect example of UFO Deja Vu— of repeated phenomena by something of "other worldly" origins, be it extraterrestrial, interdimensional, or something even more "extreme." Conventionality does NOT apply here. Not only do you have to think outside the box, but you might have to throw away the box entirely. You will note the similarities between the multiple UFO phenomena on this ranch in Colorado, to the mysteries that surround Skinwalker Ranch in Utah. You've no doubt familiarized yourself with the weird events that transpired under "scientific" conditions on what was once a large piece of property owned by the Shermans. This report by the prestigious APRO organization headed by the late husband and wife team of Jim and Coral Lorenzen will blow your mind. If you think that all UFOs are of interplanetary origin, you will soon change your mind, we can guarantee that, unless you wish to bury your head in a cosmic dust storm. Oh, and please excuse any typographical errors you might come across. There are bound to be a few glitches as the original copy was probably set in type by hand, or with a technologically primitive device known as a VeriType machine which was known to cause some skipping problems, particularly when it came to quote marks and dashes.

We have done our best to clean up the manuscript to make it more readable as the original was very closely spaced with longer than usual paragraphs. This report is almost guaranteed to make you sit on the edge of your seat, especially if you are new to the world of UFOs and its connection with fractions of the paranormal.

There is definitely a portal at play here. If anyone should know the exact location of the ranch or who the current owners might be, please contact us in confidentiality with this information – mrufo8@hotmail.com

* * * * *

This report is unusual in several ways: no names of individuals associated with the experiences are given, and no information is given about the location of the area; the individuals describing these events have not subjected themselves to polygraph examinations, personality inventories, or other methods of personal assessment. The reasons for these procedures are not because the participants are fearful, on a personal basis, of any evaluation; rather, they are concerned, on a social basis, with revealing information about the identities of persons with whom they have experienced these unusual events. They do not wish to subject their friends and families to inquiries from the general public or from military or governmental representatives.

Thus, the investigators share some responsibility in providing information about the claimed UFO experiences and in preventing information about the witnesses and locations being known to others. This request for confidentiality of information was made, and approved, on the basis it was the only way for information about these UFO experiences could be shared.

PRELIMINARY INFORMATION

Dr. Derr, who serves as a Seismologist with the U.S. Geological Survey in Denver, Colorado, received a telephone call from individuals who expressed an interest in his credentials as a scientist and as a UFO investigator. At first, the individuals did not wish to give their names and addresses; however, after initial discussions, the persons agreed to conversations in the home of John and Janet Derr, with the understanding that they were representing several persons who do not wish to have their names and addresses revealed.

PRELIMINARY INVESTIGATION

Dr. and Mrs. Derr met with the two individuals and talked with them about the events which have occurred on a ranch near a small Rocky Mountain community. For this report, we have called it ,"Clearview, Colorado." The participants in these various experiences will be identified as follows (titles and relationships are correct, but fictitious names have been used).

The Business Partner [Jim]: The Business Partner is a middle-aged man who has professional training in physical and biological sciences, and who has served in the U.S. military services, including a position as a Public Information Officer. The man is intelligent, educated, articulate, and familiar with the techniques of military operations, including military secrecy.

The Family: Husband and Wife [John and Barbara]: The Husband is in a management position in a large corporation; his position could be threatened by any undue personal publicity. The Wife is a middle-aged woman, who is intelligent,

UFOS DEJA VU

articulate, and perceptive in regard to people's feelings and attitudes. Their teenage sons share the interest and concerns of others in the community, and the oldest son (Joe) was allowed by his mother to describe events in which he had participated.

Others who have described events in which they had personal experiences are the Photographer (Roger), who photographed areas of unusual conditions; the Neighbor (Connie), a friend of the family who has known the Family and Business Partner for several years and who was willing to describe her UFO sightings; the Law Officer, who was willing to describe his sightings and his investigations of animal mutilations.

FURTHER INVESTIGATION

After the initial statements by the Business Partner and the Wife, Dr. Derr and Dr. Sprinkle continued to participate in further investigations. Dr. Sprinkle serves as Director of Counseling and Testing and Professor of Counseling Services at the University of Wyoming, Laramie, Wyoming. He has been interested in UFO investigations for many years, and serves as APRO consultant in Psychology. Because Dr. Derr has a private airplane as well as a private automobile, he is able to travel long distances in a relatively short time. Dr. Sprinkle met with Dr. and Mrs. Derr to discuss the preliminary investigation and to travel to the community to meet with the concerned individuals. Interviews were conducted with the Family and the Business Partner, the Neighbor, and the Law Officer. In each instance, the individuals requested that no publicity be given to their names and locations, although they were willing to discuss the events which have occurred and their own reactions to these events.

It was obvious to Dr. and Mrs. Derr and to Dr. Sprinkle that the individuals involved were under some stress as they described their experiences (nervousness in voices; concern about not talking while in public places; willingness to talk with trusted friends, but a wish not to discuss matters with persons who might disrupt the friendly relationships between individuals in the community). The "escalation hypothesis" to account for these strange events went from "tricks by neighbor boys" to "military operations" as possible UFO events. The evidence, according to the claims of these people, is massive that unusual events have occurred; however, the evidence provides little in the way of "proof" that these events have occurred. The significant effect is the emotional impact on these people; they all state that the events have altered their lives, and they believe that the unusual experiences are part of a huge operation. The Business Partner and the Family are willing for the investigation to continue, although they are no longer living on the ranch. The events caused them such concern that they felt they were forced to abandon their hopes and dreams for a Rocky Mountain ranch life.

UFOS DEJA VU

SUMMARY OF STATEMENTS

(For detailed information of statements, see Appendix I: Chronology of Unusual Events].

The UFO witnesses have described, in various conversations, a variety of UFO sightings and related experiences. These events are summarized as follows:

A few years ago, the Business Partner and the Husband and Wife pooled their financial resources to buy a ranch in the Rocky Mountain area. The ranch was to be renovated and established as a working cattle ranch. Despite the long distance from their original location, they were pleased with the size and features of the ranch, including adequate grazing area, woods and springs feeding a pond near the ranch house. They were puzzled because the ranch had been abandoned for several years, prior to their settling on the ranch. Also, they were puzzled because of the "unspoken" mystery surrounding a building which had "disappeared," more than a decade earlier. After settling on the ranch, they experienced a variety of "unusual events:

1).— A humming sound often was experienced in their house, although the hum could not be traced to electrical systems (which had been rewired), or to any other devices in the house. Often the humming sound was loud and distinct, but seemed to follow a pattern of being heard during a wind storm and for an hour or so after a wind storm.

2).— Noises indicated that someone was walking around outside the ranch house; at times someone seemed to open the car door and beat upon the walls or doors of the house and then run away.

3.— Creatures who looked like Big Foot, were seen in the woods.

4).— UFO sightings included nine glowing discs, which set down in the pasture near the pond; during that observation, a light flash was observed by two men, including the Business Partner at that same moment, the Wife, who was looking out of the window, experienced a "blow to the head;" she fell unconscious to the floor and was revived within a few seconds.

5).— The Photographer took a picture of a large circle (approximately 75 feet in diameter) in a clearing; initially, the "circle" supported no vegetation, but now the grass and weeds are beginning to grow again.

6).— The Business Partner observed a landed disk in the woods and saw two humanoids nearby. He and the oldest son in the family saw a box (approximately 3 feet long and 18 inches wide and 18 inches thick) which emitted strange hums and other sounds, while flashing on and off with multicolored lights, giving an appearance of an electronic device.

7).— One morning the Business Partner awoke, unable to move, and saw a tall, skinny creature wearing a helmet standing in the room by the glass door, observing him.

188

8).— There were many instances in which the electrical power of the ranch house and buildings were cut off, causing the area to be plunged into darkness; however, other ranchers did not have electrical power blackouts during these episodes,

9).— On one occasion, during an electrical blackout, the radio and stereo systems of the ranch house and bunk house emitted sounds of a voice saying, "you have been allowed to remain. Do not cause us to take action which you will regret."

10).— The Business Partner described his experience of seeing two small humanoids near a landed disk in the woods.

11).— A cattle mutilation occurred around the time of increased UFO activities, and the cow was identified as belonging to the ranch.

12).— A body of a young bull was found shortly after a hunting episode. The oldest son and his friend were frightened when they discovered the cow because they had seen either a bear or Big Foot in the area. The young bull, which did not belong to the ranch, was found with his head twisted back and with various organs and parts of his body removed as if by surgical operation.

13).— The Business Partner described his experience of being told by a humanoid that he should not go near the box; a Big Foot creature was seen walking toward the box and then falling to the ground. The humanoid said to the business partner, "As you can see, it can be lethal."

14).— The business partner described his experience of being in the woods, where he saw a disk and two humanoids, who looked and acted as if they expected him to arrive; they spoke to him in English and told him that they were sorry for the inconvenience and disturbances which had occurred to the people living on the ranch; they said that they expected some changes to occur in the events which were happening at the ranch. However, after the encounter and seeing the "skinny creature," the number of strange events, and the tension associated with the experiences, seemed to increase for the people living on the ranch.

15).— A fire began on the front porch of the ranch house, apparently started by electrical wiring in contact with paint buckets on the porch.

With the continuation of the strange experiences, and the fear and stress associated with the events, the business partner and the family decided to abandon their attempts to live at the ranch. Now, the business partner and the family are engaged in a small business operation; however, they are hopeful that someday they may be able to return to the ranch and to develop the property into a working ranch.

After talking with the family and friends of the oldest son, a conversation was held with a woman (Connie) who has been a friend of the family and a member of the community. She described a UFO sighting which occurred 12 or 13 years ago;

she also described a later sighting, which increased her puzzlement and concern. At first, she assumed that there was some military explanation for the sightings; but as she learned more about the feelings and experiences of her friends and neighbors, she began to wonder if some other explanation might account for these strange events.

The Law Officer was quite explicit in requesting that this information not be publicized; he did not wish to cause any public ridicule to be brought upon his office, and he did not wish to cause any ridicule or publicity which would disrupt the harmony of the community. He said that in 1975 there had been several hundred cases of cattle mutilations reported, and he had investigated many of these cases. In 1976, there were fewer than a hundred, and in 1977 there were fewer than ten. However, he pointed out that during 1977 he was following a policy of going out on very few calls regarding reported animal mutilations. This policy apparently was an intentional method of reducing the number of reports of cattle mutilations; however, no one knows whether it is the policy of the Sheriff, or whether it is a policy which is being followed by law officers of other counties, states, or regions. The law officer described his fear and apprehension when, out on night patrol, he saw strange lights near the road. He knew the territory well enough that, if he were to drive down the road, he would know where to find a car or truck, or a vehicle on the road.

However, many of these lights seemed to be silent objects which moved "off the road" and into the air, revealing cockpits as they flew away. He did not claim to see any humanoids associated with these lights or objects, but has viewed enough of these events to know that they are very real and that they cannot be accounted for by the usual explanations of airplanes, helicopters, hallucinations, or the "planet Venus." Although he cannot verify the events described by the Business Partner and the Family, he knows that their earlier concerns were based upon the assumption that neighbor boys were playing tricks. Later, they questioned him about the cause of the reported mutilations. Now, the law officer is aware that military operations may not be an adequate explanation for the experiences and the sightings which have occurred near the area. The Law Officer does not profess to have an explanation, but he no longer scoffs at the hypothesis of extraterrestrial visitation.

CONCLUSIONS

Although it is difficult to describe and explain the claimed events, it is quite apparent to the investigators that the people with whom we have been talking are intelligent, articulate, and perceptive individuals. The intellectual and emotional doubts (and courage) which are experienced by these witnesses are apparent in their conversations and in the way they have conducted themselves during the interviews.

UFOS DEJA VU

The UFO witnesses seem to be experiencing a dilemma in which they wish to be seen as conscientious and patriotic Americans, and yet they are convinced that "U.S. military operations" is not sufficient explanation to account for the strange events which they have experienced. Their concern about the meaning of these events is enough that they wish to share information with trusted persons, in the hopes that investigators may be able to provide them with a better explanation for these unusual events. Within the limitations of maintaining the confidentiality of witnesses' names, and addresses of witnesses, it is hoped that this report will provide APRO consultants with information about the investigation of the unusual events, and it is hoped that some suggestions can be shared on possible methods for further investigation of the area.

APPENDIX I

CHRONOLOGY OF UNUSUAL EVENTS

Note: The following is a condensed transcript of the initial interview with the Business Partner (Jim) and the Wife (Barbara)—not their real names. The conversation has been somewhat condensed for clarity and easier reading, and reordered slightly to group descriptions of the sightings and unusual events at the beginning, leaving then-speculations at the end. Some information from subsequent interviews with the photographer (Roger) and one of the frequent guests (Harry) has been inserted as if said by Jim. Jim and Barbara have reviewed this entire manuscript and approved it for release in this anonymous form.

1. The House Hums

Jim: The first unusual occurrence, after we moved onto the ranch, was a strange hum. We went out and disconnected all the electrical wires. And at first we just thought it was an unusual occurrence as we had done the wiring ourselves. But the house hummed most particularly during a wind storm and for some time afterwards. I found out later that the ranch may be undermined with a large amount of coal mines. It was one of the largest coal areas in early Colorado history. And supposedly one of the largest is on our ranch, but we've never found any traces of it at all. And I've really looked; I've walked every inch of the land. I still believe that there are some kind of pumps underground that pump the water out due to the barometric changes that raise the water level after a windstorm. It is almost invariably associated with the wind, it is quite loud and distinct; and on the occasions that we have seen UFOs, the hum usually comes along and it is quite similar. I don't believe aircraft would go up every time the wind came but the house does hum almost with every major windstorm. I'll play you a tape of the hum that we recorded. The house does this continuously during a windstorm and usually for an hour or two afterwards, depending on the length and intensity of the storm. It was humming the last time we were there—about six months ago—and I'm sure if you go there and could spend in excess of 8 hours there, you would hear it. The most practical time to hear it is about 6 or 7 o'clock in the morning, almost religiously, it is like an alarm clock. It does come on at other times during the day but it is easier to hear it at night. There is a particular area that it emanates from stronger and that is where we put the microphone to record it. If you want to make a tape of this to analyze it, feel free to do it but I don't want to give up the tape.

Barbara: It sounds almost like a turbine of some kind. The tape doesn't give a clear sound of it. It would come on sometimes louder and sometimes softer. Once, the first year, at Christmas, it came on so loud we had difficulty talking and hearing each other. That is the only time it came on that loudly. It never again came on quite that loudly. My husband was the first one that heard it. We had just finished moving the last few loads in and Jim and I were still in town picking up some more things. John had gotten to the ranch first. The children were at a friend's. The minute we walked in, John said there was something wrong with the wiring and there is this terrific hum in here. And we had put all the wiring in ourselves. We worked on the house for three months before we could even move in. There was dirt all over it 'cause doors had blown open and it was just terrible. Jim had put most of the wiring in and we thought that maybe we had done something wrong, so we shut off all the breakers we could find. And that didn't stop it so we went out and turned it off at the pole, so there was no electricity at all coming in from the pole which is in the center of the corral.

2. Bright, Trapezoidal-shaped Light

Jim: I moved to the ranch in October of 1975. When we first moved out there, we had cattle on the ranch and I was watching the cattle and remodeling. And on approximately October 16—we have a dam on the northwest corner of the property and the cattle were braying very heavily that night. I went out to see what was disturbing them. We have a large coon hound that was watching the property and the dog was extremely afraid of something. He was on the porch and wanted in and I grabbed a gun and went out. The cattle—range cattle do not usually come close to you; they usually give you a wide berth—were packing me so tight that I couldn't hardly get through the middle of the herd. I walked about halfway to the dam, and above the dam was a large lighted object. It was orange and trapezoidal shaped, and looked much like the top of a lighthouse. I was going to see what it was, and I got about halfway there when I realized that the cattle were upset, the dog was upset and that whatever it was I didn't need to know! That was during the cattle mutilation days and I had a small caliber rifle and I decided to leave well enough alone. So I went back to the house and just filed it away as something interesting that happened, and tried to forget it. Shortly after that we had several cases of where the dog wouldn't go outside. The dog is not afraid of anything except, we found out later, he's afraid of bears. We have a mountain lion on the property which the dog attacks very rapidly, so we couldn't figure out why the dog was afraid.

UFOS DEJA VU

3. Paralysis

Jim: Shortly after that, Barbara moved out there, I think, and John was commuting back and forth to his office so he wasn't there a good part of the time. We didn't have the heat on very adequately and it turned cold, and we had electric heaters set up all over the place using the electricity to heat us. We were all in one bedroom in the back watching TV. There were three of us: Steve, a friend from Boston, Barbara and I. Well, I found that they had a large reward for catching the cattle mutilators. That night I was going out tracking them. I had a large 12-gauge shotgun and I fancied myself the great hunter and I was going to go catch the cattle mutilators. Well, I laid down on the couch and I was unable to get up; it was like paralysis, like I was drugged. It was about 8:30 at night. Now, Barbara, you tell what happened to you.

Barbara: We were watching TV, Steve and I, in the bedroom that was fixed up as a sitting room. Without any outside stimulus that I know of, my blood pressure and my heartbeat went up. I do have a high blood pressure problem. I saw spots and had difficulty breathing and thought I was having a heart attack. Steve panicked and I was trying to get Jim awake on the couch because I really thought that was what was happening. I have a perceptual problem with direction. I do not know north, south, east or west. But all of a sudden I knew exactly where I was—a feeling I have never had before and have never had since then.

My thinking pattern felt different. I don't think logically. I'm a very emotional person and things that popped into my mind were just like pearls on a string and went right down to the end. I remembered things that I had completely forgotten and that frightened me, and by that time I was screaming and Jim came to from whatever his problem was—he couldn't seem to talk—and I was trying to tell him what had happened. And while I was getting it all out the only thing I wanted to do was leave. Now you have to realize that I loved that place and some part of me still does. I had no place that I wanted to go but my immediate feeling was get out, get in that car and go anywhere. And I wasn't at the point where anyone had to hold me down, but I was having to hold myself down. I really wanted to run...anywhere. And Jim got me calmed down and Steve was in a panic. And shortly after that, John showed up. I started to tell him what happened but I couldn't talk about it. Every time I started to, I would stutter. Jim then started telling him what happened, and when Jim started telling him, then I was able to talk about it. That was the first strange thing that happened to me—something like it had never happened to me before in my life. I have never lost control. Even drinking—I watch very carefully what I drink because I might do something wrong and someone might laugh. I was upset for days.

Jim: I am a late night owl and I couldn't conceive of myself going to sleep at 8:30 and not being able to get up but again; we had nothing to relate it to as being an eerie experience except as something strange happening and just forget it.

4. First Mutilation

Jim: Shortly after that, the oldest boy, Joe, had his friends out from Denver. They were out going through the woods—this was a Sunday morning—and they found our first mutilated cow about 200 yards from the house up over the hill. They came back in a panic. They were both 16 years old. It was snowing heavily and we decided the first thing to do was get into town and get the law officer. The boys were pretty upset and we took some time to talk to them 'cause they felt that something was following them to get them. I assumed it was a natural paranoia. We went out and I followed the route and we found huge footprints that had followed them all the way from the cow to the horse barn and the footprints were even in the soft manure inside the horse barn. They were, I guess, 18 inches;— I didn't measure them but they were quite large—what I guess you would call a "big foot" footprint. Then immediately, I decided to go get the law officer. I went to the nearest phone and called the law officer and told him I wanted him out right away. He said he would be out next Wednesday. I explained the footprints—he said, well, he couldn't make it.

But there was a bad snowstorm and he didn't show up the next Wednesday. When we went out the next day, all the footprints in the snow had been removed, even our own footprints. Ail of them were just gone. That was pretty upsetting and I was pretty angry with the law officer by that time. The mutilation was discovered early Sunday morning. The boys came back about 9 A.M. and the footprints were there all day Sunday, but Monday morning they were all gone. The law officer didn't show up which angered me. Wednesday, when he was supposed to show up, he still didn't show up and we still had a mutilated cow. The udders were removed surgically—the sac under the udders wasn't perforated. One eye was missing. One ear was missing. And that was it. There was no blood. All the blood had been removed. And there were no tracks in the snow around the cow. It was just there, mutilated. John states that the rectum was also removed.

Barbara: There were some funny looking markings in the snow, and we found out later they were caused by owls. We found that it took two weeks after the mutilation before any of the wild animals would touch it.

5. Second Mutilation

Jim: Fourteen days after the first mutilation, a friend of ours from California, a forest ranger, stopped to visit. We went out walking and were showing the ranch to him and other friends, including a photographer, Roger, who had been hired from Denver to take pictures of the mutilated cow. We found our second mutilation, a bull, as we were walking over the property, and the bull wasn't ours! Subsequently I went into town and called the law officer. He said he would like to talk to me, so we met at a restaurant in town. I asked him why I hadn't heard from him and why he wasn't trying to solve this. He explained to me that they knew what the mutilations were and they had known for some time. They only report one out of four, and in this county alone, there had been over 400 reports by that time. It was being done by extraterrestrials, and they had spoken to the FBI about it. I told him that I couldn't believe it. I said that I could believe that an intelligent race could travel across space but to do some of the things they had done to the cow and bull stretched my credulities a little too far. I told him that if he couldn't solve the crime, at least don't blame it on something like that. I made an enemy of the man, and I frankly didn't care because we had lost two cattle.

6. Chased by Dark Shape

Jim: The photographer asked if he could come out the next weekend and bring a friend, Harry, with him. These two and Steve, who worked for us, were standing in front of a log cabin about 10 o'clock at night when they heard a strange noise coming from the cistern which is southeast from the house and about 60 yards up the slope in the direction of the burned spot. All of a sudden, after listening to this noise, a huge, dark object pushed its way through the barbed wire and came straight at them down the hill. They all came running into the house. I went running out right away to catch whatever it was, and then I thought they were putting me on. We subsequently went back and traced the route and sure enough, there were footprints. I removed some of the hair from the fence, noticing that it just pushed its way through the barbed wire —it didn't jump the fence but just by force alone spread it and went through. There were long strands of hair —I collected the hair and got photographs of the footprints coming to the house. By then I was getting more and more upset about the activities that were building up continuously over a period of time. I sent some of the hairs down to Denver to a biogeneticist for examination. His report later was "no known species." By then I was collecting books and discovered that other people had sent in hair and nothing had ever come of it. And one law officer was encouraging me to keep my mouth shut about it because he didn't want a full-scale panic in the county. I told him I wasn't interested in panics, I was interested in finding out who was mutilating my cattle.

7. Disk Cruises by House

Barbara: It got very heavy after that. The closest you can define it is an emotional attack of some kind. Everyone went into absolute feelings of fear and depression with no stimulant. I have very tractable children; it is amazing how well they get along. But everyone was fighting with everyone. Just a lot of little paranoid incidents, one after the other.

Jim: By this time I was sleeping on the couch in front of the door with the gun and staying awake most of the night to catch whoever it was. I was beginning to suspect that somehow the real estate man was involved in it; that he was trying to make us break the contract so he could resell the land. I was lying on the couch about 2 A.M. There was a humming noise again and I raised up and looked out the window. The disk came out of the north across our property, very slowly, right in front, of the front windows and went up the gully past the whole front of the house.

Barbara: I did not see the disk—what I saw was the glow from the disk; my bedroom was on the other extreme of the house and I could see a glow over a slight hump. I screamed and. Jim came running to tell me what he had seen.

8. Friend's Car Chased by Disk

Jim: By then I was getting a little more upset and I went into town. Two of our best friends owned the restaurant and we sat down and we talked to them about it. They then opened up and told us about the property and some of the unusual occurrences. They told us about the people down the road who had just moved into the county and had mentioned an incident just prior to that. Prior to our moving out there, their cousin and nephew were going down the road past our property. A disk allegedly came off of our property, (it wasn't ours at that time) and followed the car. They went down the road, turned the corner, and pulled into their house. They ran inside and while all the people in the house watched, the disk hovered over the car for approximately 45 minutes. Later, she went into town and talked about it. After hearing this, I went back to the law officer and apologized; I had to eat crow over the fact that there obviously was something going on. We subsequently patched up the friendship and I think we are so-so friends now.

Barbara: He has been very supportive in coming out to help when we have had problems.

Jim: And he moved out there by us shortly thereafter and told me that he, frankly, was quite scared but that it was his duty. He didn't feel that he could do anything but that if we needed him he would come. This made us feel a little bit better.

9. Neighbor Has Seen Lights for Years

Jim: We found out that there was a "Crazy Lady" in the county; we had heard a lot of stories of her calling the law officers with stories of lights over our property before we moved in.

10. Apparent Harassment by Hairy Creature

Jim: During this time I felt that something was trying to scare us away. We heard a slamming noise and I went out to our car but didn't see anything. Then I saw the trunk light on the dash of the Cadillac. The trunk was locked; I opened it and saw that the pin switch had been bent over to the side. Something had opened and closed the trunk and bent it when only I had the key. I again filed that as something unusual. Something would continuously come up and beat on the sides of the house twice and run—you know, just like Halloween. Again, I decided it was the real estate man. I went down and talked to the law officer again; he requested that I not shoot anything. He was afraid that whatever it was could shoot back harder—much more undesirably than I could. So it continued progressively for some time after that beating on the walls and running. I would run out and I would see a big hairy thing running. I was tolerating it to a certain degree. One night I didn't —I went out and one was running beside the corral and I shot it. Barbara came running out and when Harry and Roger came back from town we went tracking to see if we could find what I had shot. It didn't seem too hurt it at all; there was a little flinch.

I'm a good shot so I know I hit it. There was no blood, no traces, no signs. We pursued it onto the property next to us. Then I heard this most unusual sound. They said it was like a double sound—it was a whine with almost a beeping noise intermixed. The closest sound I think I could compare to a South American primate of some type. Some people said it sounded mechanical; it didn't to me. And we roamed around looking but didn't find anything except we seemed to be led in a certain direction'. We didn't find anything; we returned to the house and I notified the law officer that I had shot one. He gave me hell and told me I was lucky this time and he didn't want anybody killed.

11. Communications

Jim: They stopped bothering us quite so bad, I think, after that. We came home one night after shopping. The three boys were alone at home. That same sound was going on to the east and one up by the barn (west) like calling back and forth. It was like they had an intelligent pattern of communication. It was broken syllables—it wasn't like animal sounds. It was like it was their communication. I joked and said, "Well, the least you can do is come down and help us carry the groceries." When we went in the three boys were hidden in the back bedroom, terrified. From what they said, something had been beating on the house continually that night. Well, I have a pretty good temper which I have lost only three or four times in my life. I stormed outside—I don't remember exactly what I said but most of it can't be repeated. I think I threatened that if we can't have the land, then you won't either—I'll blow the whole thing away. I meant it too; I would have destroyed it before I would have moved right then. I was mad. I went into the house, had some coffee and calmed down. Then, since our septic tank and toilet weren't working, I went outside.

While outside, this voice just came out of nowhere and said four words, "Dr. Jim, we accept." I think that was the first time I was really shaken rather than them just angering or disorienting me. That was all, just like a stereo, it came from everywhere. I came in the house and was pretty upset.

12. Cars Damaged

Jim: One night "It" destroyed two cars. Barbara has a green Cadillac and a station wagon. One night when we went to leave, the transmissions on both cars were inoperable. We had to have both cars repaired. Again, I assume "they "It" did it; I was furious as I never had two transmissions go out like that on two good cars. I again felt that we were being driven off and I was still trying to believe it was the real estate man. Both cars required total transmissions as all the gears were shot.

13. Shiny Black Box

Barbara: I wasn't feeling well that night and I went back to lie down in the dark for awhile. I had pulled the drape part way on the window creating a triangular area. There were some patches of snow outside and from where I was lying on the bed the triangular area was lit up by a background of snow patches. I lit a cigarette and as the match went out, I realized that the triangular area wasn't lit up anymore. I scrambled down to the foot of the bed and looked out. I could see only a middle section of something that looked like a box. It was black and very shiny. It wasn't lit but looked like it reflected light. There was a rounded shape around it. It looked like something was carrying it under his arm. It was dose to the window—within two or three feet. There is a line of trees behind the house and it was between the trees and the house—the trees are about four or five feet from the house. I ran in to tell everyone what had happened. We ran outside and, as usual, no one could find anything. This has happened more often than not. By the time you tell someone or get your coat on, etc., you couldn't find anything. But we found that if we didn't turn the porch light on, we could get out quickly enough to hear something.

14. More Harassment

Jim: There were many smaller incidents that were unnerving. A friend of mine from Boston came out but I finally had to tell him to go back. He was becoming paranoid and thought that whatever it was had come to get him; he was going to go out and let them get him because he couldn't stand waiting any longer. We had several friends come out who were harassed one way or the other or terrified; a car door opened and closed when it was locked. I went in and talked to the law officer again. If he really felt we were in any danger, I would move the boys. He said that nobody he knew of had been hurt. They had lost horses, a lot of cattle and animals of one kind or another and people had been terrified. The mutilations were going on at a much heavier rate during this period—this was 1976—the mutilations were occurring weekly.

15. Neighbor Family Terrified, Move Away

Jim: About, that time, the school bus driver, who was watching a large ranch for the owners and had some cattle of his own, moved back to Nebraska after being terrified. His son's yearling was mutilated and the boy evidently saw something that terrified him.

UFOS DEJA VU

Barbara: His mother came over and asked me if we had seen anything like her son had seen because his father hadn't believed him. He had seen something very large and was really scared. I didn't want to discuss it because of the children but said that we had some strange occurrences.

16: Plane Crashes

Jim: Also, during that period, two airline pilots and the son of one of them came out and wanted to put a landing strip on our land in exchange for my use of their plane. I agreed. About three weeks later, one of them and two others were killed in a crash nearby in clear weather. The plane was found in the daytime, but I don't know if it happened in the day or at night. Others who have inquired deeply into the mutilations, I understand, have disappeared, including the editor of a magazine who was never seen again. I also understand that two National Guard interceptors were up close to here and went down while in pursuit of a UFO at night. All of this was a little unnerving so I decided I was going to stop my own inquiry. I was going to just quietly mind my own business.

17. Nine Disks Land in Front of House

Jim: The forest ranger, David, from California, came again to visit. It was late at night, about 2 A.M. Almost all of these incidents have happened at night; we have never seen anything during the daytime. Peggy and Harry were also visiting. Nine disks landed in the front yard and I got a very clear view of them. They matched exactly one of the pictures I had seen in a book I have. Harry and Barbara were at the window and David was asleep in the bedroom. I called for David and I started out the front door. I was walking toward them as Barbara and Harry watched from the window.

Barbara: Harry and I were watching from the window, both from the same window. What I can't figure out is that Jim and I saw such a large array of them but Harry saw large, dark football shapes as if they were blocking part of the view. We were trying to see what would happen in the whole area—and watching Jim walk out, which I didn't want him to do. I have no awareness of what happened as far as what anyone else saw from that point on. My face was close to the glass and something hit me in the forehead, a forceful impact. I was knocked back off the couch and fell to the floor. Harry saw light out of the corner of his eyes, and Jim said he saw a flash of light. I didn't see it.

Jim: Harry yelled to me and I said I saw it. Harry said ,"Barbara, it's Barbara." And I went running back to the house and found her in the middle of the floor.

Barbara: The next thing I remember after looking out and seeing them was Jim putting a flashlight in my eyes to see if I had a concussion.

Jim: While I was working over Barbara on the floor, they disappeared. I thought about it subsequently and wondered why they didn't do it to me since I was the one walking toward them. But then I began to understand that they did the most practical thing that could be done to get me back inside and get both people away from the window. And I think I really began to respect how clever they were. Then I began to suspect that maybe the government was doing it to us. Except that there were a lot of unusual things. For instance David had been paralyzed during the incident; he could hear us calling but couldn't get up until it was over. But then he went out walking with me and what we call ultrasonics, the extremely high-pitched sounds, were going on out there and continued all night. David was sick then for three days. The sound would sometimes give us headaches but not all of us at once, to different individuals at different times.

18. Box With Blinking Lights

Jim: I went back to the law officer and had a talk with him and he started telling me about some of the incidents that had happened again and about how he had pictures. He said they have a box and he had seen it on occasion. He had seen blinking lights where there shouldn't be any, in trees, and such. He said he was out on a patrol one night and he saw, in a group of trees, this box that was blinking. He said he didn't want to go in alone so he raced back to town and picked up another law officer to go with him. When he got back, the trees were gone, the box was gone, everything was gone. He thought they had gone into the ground; he had seen things go into the ground before. He is reasonably convinced that they just go into the ground; I've never seen that happen.

19. Black Box Makes Angry Sounds

Jim: There is one piece of what you might call physical evidence on the ranch; there is a big burned spot on the top of the hill approximately 35 feet across where nothing would grow the first year. It's beginning to grow in a little now. On a compulsion, the older boy, Joe, and I drove up there one night and parked at the circle. In the trees, a bright yellow light, not bright I guess, a little dim but yellow, looked like an old car headlight, just shined on the car. The back was toward the trees. We got out and walked over and there was a box on the ground. I told Joe to stay back about 10 feet. It was making a buzzing sound just like zzzzzz and there was light inside it but not on it. Hard to describe. It was night but there was a full moon and as I walked to about four feet from it, it changed its tone entirely. It sounded like a bunch of angry bees. The sound went up so I backed away and I told Joe to go back to the car and watch me as I walked up to the box. We then walked back to the car and I told Joe that whatever happens, do not leave the car. Then I walked back and the box was gone. Following that is the part that Barbara prefers that I leave out.

Barbara: Absolutely!

Investigator: Are you leaving it out because it's personally embarrassing or because it's terrifying.

Jim: Too incredulous, that's the part that is too kooky, frankly. And she requested that I didn't. And I don't think that close encounters of the third kind are really interesting to anyone except to whom they happen.

Barbara: If something should happen, something should go wrong and our names should be connected with this, I could face it but I could not face this other situation. And I'm too psychologically upset—almost destroyed—by this whole thing anyway and I'm trying very hard to keep from feeling that I have experienced something that I didn't experience, or did I experience it and can't remember, or what.

(At this point, Barbara is shaking so badly she can hardly light her cigarette. She has been practically chain smoking since the beginning of the narrative.)

20. Large Bird, Dog Notes Strange Scent

Jim: There were quite a number of other things. I think they fit pretty much into a pattern. One afternoon I went out walking in the woods and I saw a bird that was about three feet tall. I got one clear side of it. It was brown and had three feather-like appendages on its head. We have a whole set of animal books and I went through them trying to track it down. I didn't find that any such animal existed on earth as I knew it. I tried to follow it and it went around those rolling hills and was gone. I came back to the house, and as I was coming through the fence, there was a slight snowstorm. We have this huge coon hound, and as I approached the fence, their kids saw me coming (they could just barely see me in the snow). The dog saw me and came running, and when he got to the fence, he stopped and started barking and growling when he got close to me. When I passed, he picked up some kind of an odor and wasn't letting me through the fence.

That scared the kids and they went running into the house, thinking that something was coming. They could just see the shape on the hill. The dog was still not letting me through the fence even though I was talking to him. It took three or four minutes of hard talking before my own dog let me through the fence. And I had been crawling all around where this unusual birdlike thing was. And I really thought about what it could be or how it could be, and I've come up with no data at all, except that it was a very unusual animal. And then it was just gone. And I don't hallucinate, you know; I mentioned that I took acid, but I've never had a conscious hallucination in my life that I am aware of. And I am sure that if I had, it would be over more than birds at the ranch, because I never had them off the ranch.

21. Buzzing Sound Comes into House

Jim: One night very late, I was lying on my couch. It was a particularly black night, no moonlight, no stars, and I told Barbara that there was not much sense in my staying awake because I couldn't see anything anyhow. I used to wait for the chickens to crow to go to sleep. It really messed up my schedule, 'cause I felt that somehow I had a duty to protect everybody. I slept with a shotgun all night. When I lay down on the couch there was just enough light remaining to see a little. I looked up, and right at the window—there was no place to stand as it was a high window—was the outlined shape of a man looking at me. He had on some tight-fitting apparel. I couldn't see any colors—just the black outline. I got up and I went to Barbara's bedroom.

Barbara: He rapped on the door. Most of what I have told you is what happened to me. This is the other thing that happened to me and me alone. I had the same thing: the blood pressure, the heart beat, the difficult breathing. I was sitting straight up in bed when Jim rapped on the door. I was trying to light a cigarette and I dropped the match on the bed in my state of terror. I opened the door and he came in and said that he had seen something strange out in front and wondered if I was alright. I told him I was having the same kind of symptoms and that I was very concerned that I was developing a heart problem. But I had gone in and had tests and they said no; I was very tense but nothing else was wrong.

Jim: She was very upset so I sat down on the bed and asked if there was anything that I could get her. I took her pulse and sat and talked to her; because when we have had really close contact at the house, it is hard to describe the feeling. People get naturally upset; they don't see anything, but as part of the pattern, I figured she would be upset when I went to the bedroom and she was. I had put a big black chair in front of the front door. We were sitting in there just talking and the front door opened and it hit the chair with a bang. And I jumped up and went running in the front room and the door was closed again. I went back in the bedroom and I sat down on the bed and I was telling her that it was just the wind or something. The next thing was this voice that came inside of my head just like a loudspeaker. It said, "We don't need to open your door to come into your home." I don't remember if that was the exact wording. And I told Barbara what I had just heard and she looked at me in a funny way.

Barbara: I thought he was losing it— I'll be honest—he knows I thought it.

Jim: And then a noise started in the front room like a buzzing noise like bees.

Barbara: It came all the way through the house to right outside my bedroom door. This I heard too.

Jim: She was holding on to my arm until she gave me black and blue spots and I frankly didn't want to get up and go into the kitchen to see what it was. I was glad she was holding on to me. And the sound just went away, then nothing more. I was sure that something was going to come into the bedroom but it didn't.

22. Huge, Cone-shaped UFO

Barbara: We were going into town very early one morning just after daybreak and just at a sharp bend in the road. . . .I think Charlie was the first one that saw it. He said, "Look at that big thing over there." I looked and Harry looked and it was cone-shaped. And I couldn't tell you the size because I don't have the ability to judge size. I didn't think of what it could be. I thought weather balloon, whatever,

UFOS DEJA VU

I didn't really pay any attention. I told Jim about it. He had gotten some books at that time. He had one that had a picture of a cone-shaped one and that's exactly what it looked like.

Jim: From the description I got from all three of them, it was immense; it was hundreds of feet, at least, across. It was six times the apparent size of a farmhouse a mile away, you know, in comparative sizes. It was a huge ship.

23. Hairy Creature Mimics Barbara

Barbara: I had only one other sighting of the animal that they described to you. I assume from what I saw later that what I saw carrying the box was the animal, only because it looked like that would be what it was. That is an assumption. We have had no less than 20 people see Big Foot at the ranch. Jim stepped out onto the porch without turning the porch light on. I was in the dining room and he rapped on the window several times without turning around and I saw him and I went to the door and opened it and came out behind him. He led me into the porch and said, "If you want to see one, I can show you where it is." And he started lining up the trees. The living room light was not on but the dining room one was casting sort of an oblique light. I leaned forward to look out and he said, "Right between those two trees, look very closely." I do not have 20/20 vision with my glasses on but I could see it. It was hunched over, and as I leaned out to look at it like this, using my hand to shade my eyes, it leaned out and went exactly like that to me. It was large and it was stooped down and it did lean down and go just exactly like I did.

Jim: The only night that we saw a large number of them was the night we picked up the kids at school. We were going back to the ranch. Do you remember when that meteorite came down over Colorado some time last year. We got a very clear view of that. It was right ahead of us on the road. It exploded and came down right in front of us. It looked to me like it hit the ranch. We got to the ranch and talked about it and heard about it on TV. I would swear it was magnesium; it flared that brightly.

24. You Can Arouse Their Curiosity: Voice from the Stereo

Jim: We have some friends from Texas who got a pretty big "jolt" at the ranch!

Barbara: They have had some very serious problems develop because they were there when a very bad thing happened.

Jim: Trust me to tell what I want to tell, will you?

Barbara: Alright.

Jim: We have been guaranteed confidentiality, the man is a professional. I know how to upset "them," and a lot of the things that scare Barbara involve their taking punitive action against us. I found that when I had guests that I wanted to see something, I could get a stack of wires and go out and get very busy with them, like I was putting something up. We would go back to the house and watch,

UFOS DEJA VU

and within a short period of time, they would be up there checking out what we did, and everyone could get a view of them. I did it several times, as sort of half a joke. Dan, the friend from Texas, was giving the old "you're not giving me that kind of junk, are you?" Electronics is his field, and he is a computer expert with a large company, and he is impressed with his own self-importance. Well, I went up deliberately to stir them up. I found that certain elements very much upset them, silver being the main element. I have a large collection of Indian silver jewelry and I discovered inadvertently that they shied away from silver.

So I went up to the circle, the burned spot I spoke about, and stuck silver bracelets in the ground along with the wires, like I was really putting something intense up. I came back to the house and we were all sitting playing Risk. I didn't know what would transpire but I hoped I would upset them. About two in the morning, the lights went out in the house, right on schedule. This voice came out of nowhere again and it was obviously intended to be terrifying and sounded like a computerized voice, very mechanical sounding. It was coming out of every radio and TV speaker in the house. We were sitting right in front of a console stereo, and the voice came out of it, and I can almost recite the words exactly. They are burned into my memory. ,"Attention, we have allowed you to remain. We have interfered with your eyes very little. Do not cause us to take action which you will regret. Your friends will be instructed to remain silent concerning us."

Barbara: That's very close.

Jim: That's just about the words. Well, Dan was extremely thrilled when the lights came back on, 'cause he said, "Now, I'm in my field." He asked if he could take apart our TV set and stereo.

Barbara: He was quite good electronically; he guaranteed that anything he took apart, he could get back together again.

Jim: And he started to dismantle it. He went through the whole unit and he said that he couldn't figure it out, but that his technicians in Texas, when he got back, would. He checked and the stereo was off; the phonograph was on when the lights went off but the radio receiver part was off; it was on phono. We found out that the type of transmitter it would take, from even close range, to cause a signal of that intensity to go through the house would be beyond our means to ever put up.

Barbara: But Dan was still sure that it was a hoax. His wife and children were all upset and crying, and Dan took his daughter off into another room and told her that they needed to find out if it was a trick or not. He told her to go back out and tell us she was frightened and wanted to leave immediately. He thought that if she acted upset enough and if we were playing a trick on them; rather than ruin the whole vacation, we would admit it. I got very upset and said that they couldn't start back to Texas in the middle of the night and I would call a friend in town to see if they could stay there. Then Dan talked to his daughter and they all calmed down and they stayed. And he just told us this weekend that he knew if it was going to ruin the whole vacation, and if it had been a trick, we would have admitted it to him. He had decided, to save his own sanity, that it was somebody else with a massive capacity for pulling a hoax on us, and that we were gullible enough to go for it. Needless to say, Dan never found the trick. He went over the whole house; he even dumped the laundry bags. He went over everything. I was glad the house was clean.

25. Don't Trust Senses Unless Two or More See

Barbara: Really, we didn't mention the fact that, often, when the wall pounding was going on or when there would be a larger number of disks or the animal would be around more; quite often all the electricity would go off. I absolutely freaked out if anybody wanted to go out and check the breaker on the corral until after things had calmed down. We would check house breakers and eventually, go out to the corral and the breakers would be on out there. We developed a system of nobody going by themselves and nobody making a big issue out of something that at least two people didn't see or hear. Because it was too easy to get paranoid. Once I went out the back door and heard this horrible sound in the corral and screamed and freaked out and came running in. Everybody went out with flashlights and it was a cow that had gotten trapped in the corral. It's very easy to get into that and I could see my children doing it. Everything strange that happened— a sudden windstorm that came up, a sudden fog that would roll in— and there would be the space creature. It was becoming an absolutely paranoid thing. It was very frightening.

Jim: I think we suffered as heavy on the mutilations as anyone I had heard of. We lost six cattle in two years. That is a pretty heavy amount of loss. In light of the fact that I was watching that land so carefully, I was determined to catch whoever it was. The reward was so high; I wouldn't have minded it at all. It now comes close to half a million dollars. The paranoia had gone down now—two years ago in the county, you didn't dare stop on the side of the road. Those people were carrying high-powered weapons and they would shoot anything that moved. It was really tense, and I can see the law officer's point of view, but I don't think you solve it by sweeping it under the rug when it is continuing to go on. The mutilations haven't gone down at all. Remember when the big mutilation thing was going on? Well, they haven't gone down at all. Investigator: When you have a problem you don't know how to deal with, sweeping it under the rug isn't good, but what do you do instead?

Barbara: I wish someone would come up with a neat package to answer that.

26. Close Encounter III

Jim: Will you remove me from my promise?

UFOS DEJA VU

Barbara: Oh, Jim!

Jim: Trust me.

Barbara: Alright.

Jim: Because I want to get it out of my mind too and then forget it.

Barbara: Alright, go ahead.

Jim: Because it's necessary to develop what bothers me. Well, the night that we saw the box, I stopped at the top of the hill and looked down into the trees and there was a light in the trees. I told Joe to go on to the house, and I walked down into the trees, and I think that's the closest I ever came to being afraid. I didn't feel fear, in that sense, but my legs wouldn't move. I had to force my legs to take me down 'cause I didn't know what I would see. I walked down to the light and there were two individuals waiting for me in the light. The light didn't come from any-where—I can't describe it—it was just light. They obviously weren't nervous and as soon as I walked up, they spoke to me by name and told me. . . . .I can quote that exactly, "How nice of you to come." It was just as though I had been expected. Down below, possibly 50 to 60 feet from us, was a disk on the ground. It was lightly lit, just light enough to see; I can describe it exactly. I've burned that in my memory. I was up there maybe five minutes; they apologized for the inconveniences they had caused us, told us that a more equitable arrangement would be worked out between us, whatever that means. I wanted to ask a lot of questions but found that I didn't—you know, like, where are you from? I didn't ask any of that.

There are several things they asked me not to repeat that have no significant meaning at all; they are unrelated to anything. I think maybe they were just check-ing to see if I would keep my mouth shut. I told them that if they were mutilating cattle, it was very foolish to do so and draw that much attention to themselves. I complained about the damage to the cars; they never admitted doing any of it. One thing they did do was that they mentioned the box and that I did the right thing backing away from it. That was what I called an implied threat. They nod-ded, and approximately 20 to 30 feet away, Big Foot, as I called him, got up and walked toward the box. The box changed tone and he dropped down.

They then said, "As you can see, they are quite lethal." They said that they would come back and talk again. There were no good-byes; I just somehow felt it was time to go. They did tell me that my memory wouldn't be tampered with. I think that is about it. I didn't ask any of the questions that I had figured I would want to ask. Somehow, they seem juvenile. And I had no doubts that these were two men. I can describe them almost exactly. I had seen them before; this is the thing I hadn't mentioned. I hadn't gotten a really close look but the two that spoke to me were not identically the same as those that I had seen before. They were similar; these were definitely humanoid. They were approximately 5 ft., 6 in. tall, I would say. They had on tight-fitting clothing, you know, like a flight suit. I noticed the clothing changed colors, -from brown to silver, but I don't know how. They were very fair, had large eyes and seemed perfectly normal, completely relaxed. They had blond hair with something over the head but I could still see their hair. They had something like a flight suit on, skin-fitted. The hair was obviously blond and wasn't long; it didn't make much of an impression. The thing that impressed me the most was the eyes, and if I were judging what they were, I would say they were humanoids. They were different than people but not different enough that you couldn't call them people.

Investigator: If you saw them on the street, you would stare at them as being different?

Jim: Right, but not freaked out by them. Their facial features were finer, their eyes were larger; they would have been striking but..... effeminate, almost delicately effeminate, completely self-assured; they obviously were handling the situation with me very well.

Investigator: Did you turn around and walkaway from them or did they go first?

Jim: I went first. We talked; there were no good-byes. It was just like, well, we're finished, and I just walked off. I thought about all of the things I would have liked to have asked but I couldn't figure out why. Then I couldn't figure out why they had even bothered to talk to me. It was obvious that I was supposed to come. They didn't say anything that would indicate why, except a more equitable arrangement.

Barbara: You weren't feeling well that night, I remember.

Jim: I was feeling very badly.

Barbara: Jim has a heart condition too.

Jim: A myocardial infarction. I didn't particularly want to go up the hill but I felt somehow compelled to go up. Nothing that happened was phenomenal; I can't figure why or how. They didn't give me any earth-shattering information or even admit they were mutilating the cattle. The only thing I found out for sure is that this big fuzzy thing, Big Foot, obeys the commands. I found that out. I found out the box can be lethal, if they were telling me the truth. It was, all in all, a very pleasant conversation we had, no trouble with them after that. This happened approximately in January of 1977. The part that was interesting was that they would see us again, and I was really excited. I came back and told everybody that they would be down to the house to visit one day. It was a very pleasant conversation and I would define them as diplomats. They were very capable of handling what they had to— they were very smooth and if I were judging by the ones that I have seen before, they were larger and they were more humanoid; if anything, they were-half-breed. They looked enough like people that in a laboratory, we could produce people that looked just like it. That was my first thought—that somehow the government was trying to do this. They were completely self-assured; they spoke vernacular English. I was pretty rocked, because I did see the disk and it was quite clear. I walked on back to the house; it wasn't very long that I was gone, I'm sure. I wasn't with them very long. I was excited over the more equitable arrangement; I guess I had some illusion that they were going to give me the cure for cancer or a billion dollars or something—at least pay for the car's transmissions. Shortly after that is when Barbara saw the other type of UFO—the ice cream cone-shaped one.

27. Tall Creature With Helmet

Jim: I was asleep on the couch; John was there because it was a weekend. It was about two in the morning. I sleep very soundly, as a rule. I woke up completely awake-wide awake and I couldn't move. I was lying on the couch looking out—there are French doors in front of it. I couldn't talk but I could breathe alright and I wanted Barbara and John to get in here and turn the lights off and see it. I was forcing the air out of my larynx and making strange sounds. They could hear me but they weren't coming. And this thing was just looking at me. And I can describe it vividly; all that was working was my eyes. I couldn't move. It was approximately seven feet tall, very skinny arms and legs, extremely skinny. It had an object on its chest, I could see the shaping of it very clearly, like a box, but it wasn't flat. It was pointed. It had three hoses on each side; this creature had a thing over its head, like a space helmet with a plastic covering. It wasn't at all terrifying; it was more or less pathetic in appearance, almost helplessly pathetic. It was just looking at me in the same way that you would look at a patient on a table, not cruelly or indifferently, just looking. I kept making these noises and it just vanished. It just wasn't there anymore and I said, "Oh, God, I'm hallucinating. I've lost my mind." Then I decided, no, I really couldn't be.

Barbara: John and I got in there just after it had disappeared so we didn't see it. The reason it took us so long was that John could not get me awake, and he was torn between running to see what was happening and trying to wake me. And we lost a few seconds that way. By the time we finally got in there, it was all over. John has had some experiences on his own and I'll leave that to him.

Jim: I think the reason that it is all so interesting to me is that we were headed toward a more amiable relationship with them, you know, after my talking with them. The disasters had stopped, the pounding on the house had stopped, the terrorism had stopped, and after talking, I kind of liked them. They were pleasant and whatever they were... I hadn't decided they came from space and I'm still not sure of that. But then again, after that, the hostilities started up again. That was extremely disorienting. The situation got extremely tense with no apparent reason. No disaster happened after that, but from the time that I talked with whatever it was on the hill until I saw the thing at the couch, everything was running so smoothly.

28. Decision to Leave

Jim: It was almost exciting that we could live peacefully with whatever it was from wherever they were from.

Barbara: I think this is what finally broke me because everything was going so peacefully and I thought we were going to be able to stay. And I really love that place and I thought everything was going to smooth out, and then it didn't.

Jim: Then after whatever it was—it obviously wasn't humanoid; it wasn't a humanoid form at all—it wasn't hostile, wasn't threatening, it wasn't dangerous. After that everything went back to double doses of tension. It got much worse, the tension, not necessarily the activity. It was a thing of... we knew we were unwanted. It's a gut-level feeling that's hard to describe exactly. We knew that something wanted us out. Barbara felt the same thing. Shortly after this sighting, we had an accidental fire with paint on the porch. It had nothing to do with them, but on top of all this feeling, that was it.

Barbara: I've often read about what they call the "Fight or Flight," and I've often wondered which one I am; well, I've decided I'm definitely flight. The only reason I didn't leave right then was that the children were there and Jim was there and how can you leave someone. But it took me an instant to make that decision; it wasn't a gut-level decision. I froze, instantly, and then I very stupidly grabbed a candlestick and ran out to the porch. Naturally what he needed was water but I didn't know that. I really thought we were being attacked. I just decided that I couldn't take that anymore because I had faced that fact that if I ran out there, I would probably die. And I figured that I was getting far off the end of the stick when going out there to die didn't seem that important. And I thought, it's time to leave before you lose it all.

29. Friend Loses Physical Control

Jim: This leads us back to another incident. A friend of a friend who was in the Army came out to the ranch to visit. He knew nothing about it; this was just his trip to the country. He spent the night but he wouldn't go into the woods; he felt something was very wrong and he didn't want to go out. We didn't press him and we didn't discuss anything. The next morning when we got up, he was already up and was walking across the fields. He would walk stiltedly out and then turn and run back; he was doing that back and forth and everyone thought he was crazy. When we asked him what was going on, he said that every time he got near the house, something took control of him and forced him to walk back into the fields.

APPENDIX O

WITNESS SPECULATION AND INFERENCE

Decision to Tell APRO Investigator

Investigator: I can see you are really concerned, but what made you finally decide to take action?

UFOS DEJA VU

Jim: I felt the events that had occurred at the ranch were significant enough so that someone of serious intent should be looking into it. Because I'm reasonably sure that there is a permanent installation there. I could go into a lot of reasons, I suppose. But the main reason is that our ranch overlooks a military installation; we have a perfect view. That is the only reason I can think of for a permanent installation being there.

I've read Hynek's book. I inquired into Dr. Condon. I know people who knew Dr. Condon. I checked into his character. I found out that the Colorado University project was basically a sham—at least in my own opinion, and that he was certainly not someone to whom I would have wanted to pass on my findings. I checked into the Herbert Shermer affair, the state trooper, and I found out how he had been treated. I knew how we treated him because I was additionally a PIO officer for the Air Force and I knew how I dealt with that sort of thing as security officer. No one with good judgement wants to be made a laughing stock.

Barbara: When Jim got to the point that he was going to write the letter, I talked him out of it. But I've known Jim's family for a long time and I know Jim well enough to know that if he wants to do something he'll go ahead and do it anyway.

Jim: I had already made plans to go to Northwestern and talk to Dr. Hynek.

Barbara: I was reasonably sure you would do something like that. Jim picked up a copy of the Sentinel newspaper and was reading the article and saw your name. We discussed it, and I called you. Whether anyone believes us or not, as we have already told you, we decided it doesn't matter. What we have seen, whether it was valid or not, someone else with some interest other than publicity should know. I'm not community-minded.

I know the areas that I'm a humanitarian in and that's not one of them. I would let George do it. I've been able to talk to some people who have really helped me understand the fact that my own fears are my own fears, and they have nothing to do with anything else, and I should not confuse them. I have gotten some of that worked out. But, anyway, with Jim bringing it up in front of everyone, I knew that he was going to do something anyway, and I felt that I would rather be in on it and know what was said and what I might have to deal with later. The thing I fear the most is the unknown. If I have a friend I have a cross word with, and I don't know how she is going to take it afterward, it drives me bananas. The unknown I can't deal with—I'd rather have the fight and get it over with. I agreed to call you because Jim is the one who wanted to give this information. I really just wanted to be away from it. John said that if you were interested in hearing what he had to say (he was at the ranch less than any of us because he had to commute so much), he would talk with you if you wanted to follow up. But he is working tonight and couldn't come. John is as nervous as I am about this because he has been with his company 19 years. He is in management and is doing well, and anything like this would

UFOS DEJA VU

totally destroy his opportunity for further advancement.

Jim: You see, the reason I decided to talk to you is your credentials. If I approach people in the town with your credentials, you know, guaranteeing confidentiality, they would talk to you. But if someone showed up with long hair, looking bizarre, and wanting to talk to space creatures; we would be pretty well ostracized, because they are tremendously clanish in the county. And there's only one place that serves hard liquor in the whole county and that's only a recent occurrence. And I really love the county. I do want to move back there but never back to the ranch.

Personal Feelings

I go back periodically. We had some guests from California and they wanted to go out and see what was happening. And I took them out and we spent the night at the ranch and we were looked over again. The reason that I think that I really wanted to bring it to someone's attention is I'm reasonably sure that they play rough. It's not big brothers from space who are interested in us as spiritual beings or whatever. I'm absolutely convinced that they couldn't care less if we live or die. We're nuisances, although I think they may be more humanitarian than we are. And I can only assume that they are watching us, watching our military potential, because I didn't conceive of anything else. I have no doubts but what they are mutilating the cattle, none at all. This cattle thing scares me a little. Certainly they have behaved better than man would have under the circumstances. If he wanted something, he would have taken it. But I'm not at all sure that their purposes and intents are at all favorable to us, or that there is anything we can do, but, at least knowledge for knowledge, it is valuable. I have no idea that there is any way that man could stop them or even impede them. But I know that they have no difficulty at all in immobilizing a person because I've been paralyzed and that's my freakout. I'm a little bit claustrophobic and when I can't move. . . with Barbara, it's her mind. I don't care about my mind —they can go through it all they want to. But don't stop me from moving. That happened about six times to me after that.

Barbara: When I was about 18, a friend of mine attempted hypnosis, and I felt the going under, and it terrified me beyond anything other than what has happened in the last couple of years. I suddenly realized that is my big fear—losing control of my mind. It is very frightening to me, losing the ability to think.

Jim: During the 60's I took LSD (it was legal) and I don't fear losing control of my mind. I find it almost enjoyable: it isn't terrifying to me at all. This paralysis has happened to several people and I can't conceive of any purpose they could have out there other than to create terror; maybe again, I think in a military way. And that doesn't go, in my thinking, with wanting to make a favorable association with man. It isn't the basis of an amiable relationship to start off instilling terror and I'm sure what they have done could serve no other purpose. They terrorized us, they terrorized others, and mutilated cattle being found all over nine states, I understand. I'm not at all sympathetic toward them, frankly. The things that went on out there left few doubts that they appear extraterrestrial, and I have few doubts that they are not friendly.

Barbara: All I want is a simple, uncomplicated life. If I never hear from you again, it will be alright. I'm not trying to discourage you but that would be fine. I care not what you do other than don't involve me. That's really all I care about... being lifted into the air, they are being drained of blood, they are being mutilated, and they are being lowered. If they wanted to do just biological research on cattle, they could have disposed of the remains without them being found. And they are left where they will be found. It is obviously some intent to instill fear and it has been quite successful. The people are extremely fearful. And, about the story that helicopters are doing it; I figured out early in the game that the government is sending in helicopters in large numbers from several sources but they are doing it to cover what is really happening. I'm absolutely sure that the helicopters have nothing to do with the mutilations. They have had intensive radar nets over that area and the law officer has been kept only moderately informed about the reasons for the mutilations.

Jim: I'm just the opposite. If you ever found out anything, I'd love to know. I'm pretty sure that the things that happened out there are significant enough, at least, like the box. I haven't read about the box but enough people have seen it. I'm reasonably sure that there is one permanent installation that can be tracked down or at least surveyed by someone. And I'm reasonably sure that the activity is increasing, not diminishing. I have some curiosity about how often this is happening. I have read reports about the alleged kidnapping of Travis Walton in Arizona. The same type of thing occurred with the man on Mt. Evans who said something was after him. And it just quietly disappeared out of the news. And of course, I'm familiar with the Betty and Barney Hill thing. I have mixed feelings—like the Hickson-Parker case in Pascagoula, Mississippi.

Barbara: I'm skeptical about anything I haven't experienced myself.

Jim: But when it happens to you, your skepticism goes very quickly, you know. It's a hard thing. I can't believe it happened to somebody else; but at the same time, I don't doubt that it happened to me. I think it is probably the same spot that a lot of people are in.

Barbara Thinks Children Have Unhealthy Interest in UFOs

Barbara: I don't want my kids to get anymore into this. The youngest boy is very bright and is getting too wrapped up in this. It is too much; it's unhealthy—too much of anything is unhealthy, I think. I don't see anything other than the tremendous, what I consider, unhealthy interest. If something comes on TV or if there is an article in a magazine, Sam will go to great lengths to hear it including sneak-

UFOS DEJA VU

ing, lying or whatever in order to get to the TV, spending lunch money for a magazine, and seeing Close Encounters of the Third Kind. And I think that's unhealthy to get that wrapped in it.

Sweet Smell

Jim: There is something I would like to know from you. When you are mining beryllium it gives off a sweet smell. I would really like to know exactly what type of a sweet smell. What we have experienced out there and a number of people have, is a very heavy permeating odor of, at first we thought vanilla, and then we finally refined it to cherry-vanilla and it permeated whole large areas. We have driven through it in the car and stopped, part way through, and it's still there. It's there during times when there has been wind which would have blown anything away from a natural herb or flower or whatever.

Barbara: A friend of my husband's who is a rock hound had found some rocks out there that he said indicated beryl possibly on the property. I don't really know that much about it.

Jim: A friend of mine was doing some library research and all she was able to find was the fact that mining beryllium gives off a sweet smell, but never found out exactly what. And there are a lot of different types of sweet smells. And I would be interested in knowing about that. I'm sure a mining engineer could tell but I don't happen to know any.

Investigator: Your ranch is located in an area of coal mines, so I think it more likely that the sweet smell comes from methane gas escaping.

Numerous People See Disks

Jim: Of course, I've insisted that everyone else stay away. Part of the reason that I got so involved, I guess I have a macho complex. I felt it was my duty to protect everybody. We saw disks very often, very regularly. John saw them, I saw them, other people saw them. A friend of ours from Wisconsin saw them. He came back one night and one of them came right in front of his car. Harry was with him.

Barbara: There were, within the two years we were there, an awfully large number of sightings.

Jim: From what I can determine, the activity has accelerated since we moved out, not diminished. We understand that from the neighbors. I took friends out there that came from California. We weren't any sooner camped—we put out sleeping bags in the woods to be right in the midst of it—than the sounds started around us. Robert has his dog with him; the dog crept up and climbed under the covers with him. He got the full dose of what we call the "creepy's"—that tremendous sense of personal danger without any stimulus. It's like a biochemical reaction. You can give injections and cause the same reaction in a person.

Coping with the Experience

Barbara: I must be honest; I have to consult a psychiatrist on my own. I have a great deal of difficulty dealing with this. I know what my own eyes saw; it isn't that my vision is poor and I thought I saw what I saw. I know what I experienced; the increased pulse rate and all that were not caused by an outside stimulus because nothing had happened to cause it. I heard no sound and saw nothing. My hearing is not that fantastic; my eyesight is not very good at all. I know there are people who have very good hearing ranges and often hear things, and subconsciously they come to conclusions consciously. I don't have that range of hearing to do that. I know what happened to me and I'm just trying to put it into some sort of reference so that I can deal with it.

Jim: I think that probably what she is saying, and I experienced it too, is the total sense of helplessness. I know what gave me the feeling was the inability to move. I never feared them; I never feared Big Foot or any of them in the sense of personal danger. But I don't think people can easily face a sense of total helplessness. I think the way I saved my sanity because it didn't really bother me appreciably, was by doing something, like sleeping with my shotgun. And I just wiped out the fact that I couldn't move if they didn't want me to. Somehow, I still felt the power to do something.

Barbara: But I don't have the range of experience that you have either. You have lived a very full life and I have not.

Jim: Well, this is not something I have lived before.

Barbara: No, not that but you have a broader base to face something with. I don't. I have never had anything that startling in my life happen to me.

Jim: Is this all absolutely confidential?

Investigator: Absolutely.

Jim: Well, what she meant by my range of experience. . . .I was a security officer in the Air Force and I transferred to a Security Agency. I worked for them when I was young and they paid my way to college. I have kind of thrived on intrigue, to some extent. So I could deal very well with intrigue. I think in a way, it prepared me to deal with the ranch. But I never thought anything like this would happen to me. And some of the things that happened, I really would like to tell you; it would be beneficial to me to get it off my chest. But I'm locked into a situational thing.

Barbara: Well, let me work a few more times with my psychiatrist and maybe we can come up with something. Since I've been trying to deal with this, my weight has gone up about 30 pounds and I'm now smoking almost three packs of cigarettes a day. Five years ago I didn't smoke at all. I'm turning myself into an absolute physical wreck and I do have to deal with it because I don't want my family to suffer. I started with the psychiatrist about three months ago. I'm sure he doesn't believe a thing I'm saying, but he is helping me at least say, alright, if you believe it, that's fine; let's deal with it from that point.

UFOS DEJA VU

Jim: She really didn't crack until that night when she thought we were under attack, when the paint on the porch caught fire; then she cracked.

Barbara: I don't know if you have ever wanted anything so bad you can taste it, as the expression goes; and that's how badly I wanted that ranch. But I couldn't face it. I think the psychiatrist thinks he is helping me. I'm sure that he doesn't believe me. But he is approaching it with the fact that if I believe it, I don't have to be insane to believe it. If I believe I saw it, and I can deal with it on a rational basis, then we can work from there. That's what I've gotten from him. You have to understand that once you face the fact, then it really doesn't make any difference whether someone believes you or not.

Two Types of Creatures Appear in Conflict

Jim: think the reason that I need to give my opinion is that these creatures—whatever they are—the humanoid ones, the ones we have seen, with the exception of the two "more humanoid" ones, have always appeared to be afraid of something. They are extremely nervous, extremely jumpy, extremely terrified of something and I'm sure it isn't us. That I have no doubt of; it isn't people. I'm sure they are watching the military base for some unknown reason. I can't think of any other reason for them to be there.

Whatever this other thing was that showed up; I actually feel more friendly toward this non-humanoid form than I do toward the ones that look humanoid. I'm reasonably sure that the humanoids are afraid of them; again supposition. Yes, supposition, you know body language, the way it looked to me; it obviously wasn't afraid of anything, if that makes sense. It was there, it intended to be there, it was almost like a compassionate thing; as if you were describing it as a religious experience, almost. You know, you come in contact with something very great. But I never felt that with the disks and whoever was on them. They had always been very nervous when anyone showed up. It was almost an extreme paranoia except for the one time I mentioned that I walked up and talked to them. They were very calm, very in control of themselves on that occasion. When I was talking to the law officer, he said that activity ceased when ships of this other type, like Barbara saw, showed up. She wasn't the only one to see it; other people in the community saw it too. And what he was relating to me was the fact that the activity would somehow go down because this other ship showed up.

There has been a consistent pattern of the disks diminishing when this other ship shows up. He didn't know anything about occupants. And again, supposition; I'm reasonable assured that the humanoids, or whatever you want to call them, are in the disks and this other skinny, non-humanoid type is in the ice cream cone-shaped thing. And what they're up to or what they are doing or the rest of it is, I think, partly what I am interested in. There is something going on between them, because I know, at the ranch, they apparently weren't at all interested in us. We were just nuisances.

Barbara: You have an opinion about what happened to me. You felt that they made some kind of mental contact with me to impress me. I know that the first time it happened, my mental pattern was changed and I wasn't even aware of it. I didn't really realize that people have that clear a pattern of thinking until it was different for me. It was totally alien to me to think that way; I don't do that, I go very much on the emotional gut level. Somebody smiles right so they must be alright. It's not what they do or the facts you get about them. And that was definitely at least a two-hour experience.

More Speculation

Jim: I think I am extremely interested in how they can get into a person's mind, because they have no difficulty controlling mind or body. Everybody has had the same feelings. Several people have seen the humanoids. Again, I was the only one who saw this skinny thing. And Barbara, I don't think she wants people to think that I am crazy. She made me absolutely promise never to tell anybody.

Barbara: The only reason he convinced me to do this was the validity of what really happened. I know what happened, I know what I saw, I know that I did see disks, I know the largest number that I saw at one time was nine, and I know that at least once every two or three months we saw disks. So I know that I saw them; I wasn't imagining them because at that time I certainly wasn't in the mental state I'm in now. I was very happy with being out there. It got progressively worse because of what happened and the only reason I agreed to do this was because Jim felt that maybe something would come of it; that maybe someone else had seen or experienced it and maybe something could be put together that would be helpful to mankind. I don't mean to sound like I don't care what happens to mankind but I don't think I could help that much. I really don't question what Jim saw.

Jim: What happened could have been hallucinatory, it could have been drug-induced, it could have been a lot of things. But from my own subjective range, it happened to me. I wouldn't want to be put up as a laughing stock in National Enquirer. That would infuriate me. But to have someone quietly think I'm insane; you know, that's their problem. Barbara wanted to hush it up, but to me, it's interesting. After the pilot crashed and other things occurred, I had no doubt that they can play rough. And the law officer encouraged me to accept that point of view. But again, if they are playing rough, they are up to something that I don't feel is ethical. I personally feel that something a whole lot more is going on. I have read a lot of books and I personally believe that Herbert Shermer, the State Trooper, was picked up. I have no doubts of that one in my mind. I have a few doubts about the Betty, Barney Hill case; I have doubts about the one in Mississippi, Hickson and Parker. I think reading the stories helped me so that I wasn't quite so upset about seeing things. It's hard though to explain what you feel inside because you never feel one solid emotion; you know, it's one over here and one over here and one over here. But I do want absolute confidentiality. That would be the one thing I couldn't deal with. If I did find that I was hallucinating; that wouldn't bother me. But I do think there is a whole lot going on with a lot of people that that's the reason I decided to come and tell it all. Some could tell a whole lot of things and I think the people that have had some of the best experiences won't. I think it is something that people should do, even if it is a subjective experience. It is possible it could have been a totally subjective experience; even with all of us, it could have been hallucinations. It could have been a lot of things; maybe none of it happened. Maybe everybody there was deceived in some way. I can't picture how but just cause I can't picture it, doesn't mean it couldn't happen.

Investigator: There is so much smoke there that it's hard to think that there isn't some fire generating it.

Barbara: I think that if you are ever going to do something material about it, that hum would be something that is still going on. To my knowledge, it has never ceased.

Jim: If you are interested in the county and the people who live out there, I'm sure I could induce them to tell about this thing. The law officer, I'm sure, will talk, but again confidentially. The "kooky" lady, Roberta, she'd talk on whatever basis.

Barbara: I met her briefly in town twice, and because of what was happening, I desperately wanted to go and hear about what she had seen. But other friends said it would not be good to become friends with her because she was "the county gossip"— there is one in every area—and she is a good, kindhearted person. She is the kind who will show up with a cake when you need it but in the meantime she will describe everything in detail that she came across. I felt I didn't need that. I have never spoken to her about anything that happened at all. But other people have told me what she has seen and she had seen it for many years before we purchased the property.

Jim: I think the only other person I have talked to, who has had as many experiences as I have, is the law officer. And, if he'll open up, he's had some lulu's. But again, what he has told me, has been in confidence. The things I have repeated, he has told in front of other people. Roger has said that he would tell what he has seen and Harry said he would tell what he has seen. Again, I feel that each person has to do what he thinks is right. We still have all the pictures that Roger took; he took pictures of the footprints and pictures of the cattle that were mutilated. I told him that for the sake of the county these cannot be released to anybody. We subsequently became friends so I trust him. We could never live there if the pictures were released. I would release them to you or a group, in confidence. But I would never let them release them to the media. That's one of the things that we have lived in absolute, dread fear of; that someone that came out there who was inexperienced, would go and tell someone else and then say, let's all go out and see.

Investigator: I'm trying to think in my own mind; where do we go from here? I think this could very well be a unique situation. I'm not familiar with anything else as extensive over a period of time. There were so many sightings.

Barbara: That's because we stayed there.

Investigator: Right-exactly. Not only does it appear to be the most comprehensive set of sightings that I know of, but also there's something still going on so that it is an excellent opportunity to investigate. It cries out for an investigation with some sort of instrumentation.

Jim: This has been my interest.

Investigator: What could we do? One of the thoughts I had was using some seismic exploration techniques to find the mine. From what you say, stringing a bunch of wires on the ground could be a very interesting thing!

Jim: It usually gets a very good reaction especially a stranger showing up doing it. I'm sure that if a person really desired an encounter, they could quickly facilitate one. I believe that very strongly. One other thing has happened a number of times to people who didn't know what was happening and came out to the ranch for the first time. After being there a few hours, they would say, "There is something very strange about this; I feel very uncomfortable." Somehow, you feel that if all these people tell you, it validates it somewhat.

Barbara: That's the point that Jim got across to me; that if you talked to that many people, you could piece together a lot of validity, and from a lot of points of view. That I do understand; no two people see something exactly the same.

Jim: Like the law officer; you can tell from talking to him that he is a no-nonsense law-officer type. I would rate his credibility extremely high.

Each person has his own level of credibility. Roberta's, for example, is very low; she'll see Venus or airplane lights and think its a UFO. Some people, like the law officer, know a whole lot more than they even told me. And I think he would tell a great deal of it, if he could feel safe doing so. He doesn't want to lose his job and he doesn't want to be blackballed in the county. But from what I picked up from him, he wants to talk to somebody about it. He's very upset and he wants to have it settled. They requested that the FBI come in and one FBI agent allegedly went out and talked to them. And it just made it worse; nobody really wants to know what is happening.

Barbara: When they had one mutilation, the law officer cut a strip himself with a knife and sent it to the CBI; and they sent a report back, saying it had been done by wild animals. I didn't hear him say this but this is what we were told. He felt that he was being given a lot of wrong information and he wanted to see if this was true.

UFOS DEJA VU

Jim: There are a lot of things I don't think he would tell me, for the same reason that I hesitated to tell you; particularly after I laughed at him once. But I think with someone he felt comfortable and secure with, he would tell a whole lot more-because he indicated to me that a whole lot more things happened. And a lot of these I can't relate; again I'm in the position where I can't tell what he hasn't given me permission to tell. But there is a whole lot more, and I feel there is more even than he told me, because he alone was doing the police work in the county. He has personally seen the ships going into the ground, he has had people tell him that they saw mutilations being done that he wouldn't talk about, who was doing it and why, and a lot of other things he just clammed up on.

TRUST

Barbara: Remember the people who lived over on that back section? He had some really strange experiences, too. They had a mutilation, like about 50 yards from their house in full view of the kitchen window and she had been in the kitchen within the time-span that it had to have happened and had never seen anything, - and this was in the daylight, early-morning hours. That happens quite often. I don't know how much you have read about the mutilations, but quite often it is in plain sight of the road; they try to track it down and find that somebody went by at 2 o'clock and somebody went by at 4 o'clock and at 2 it wasn't there and at 4 it was. That's not unusual. But anyway, he came over one day to talk to us and to ask us some questions and said that his father wanted to put up a big radar thing but he couldn't get permission to do it. I never heard what came of that; I know they moved.

Tall creature with the helmet

Jim: I think that's the thing; unless somebody opens up and invites someone to come in, a full investigation would be impossible. I told Barbara that we lived in the county and were just as guilty as the rest by putting the lid on everything. Nobody would talk and we were accessories because we were afraid of the same sort of ostracism that everyone was.

Barbara: I think if you could get the point across to your UFO investigators that there is a need for absolute confidentiality, maybe you could find out something. If it had only happened to Jim and me, we would never be here.

Jim: If an investigation was done by a professional group in the interests of protecting the United States, those people would absolutely cooperate. It would have to be approached from that basis, that somehow they were doing something good. I think on that basis, almost everyone would step forward and tell what they know if it was an accepted thing to do. I think if the law officer as a few of the other people were approached on the basis of confidentiality, on the basis that it is the right thing to do, and that it will never stigmatize them; I believe they would come forward. I really believe that. Don't you?

219

Barbara: I think a good many of them would.

Jim: The investigation would never leak to the outside. Everyone would know that it was going on, but as far as it leaving the county; it would be the same thing that somehow they would be committing an offense by telling. They have a protective thing to each other; if someone was interviewed, no one would tell. Part of the reason that I have told you everything that has happened to me is that, in order to get the law officer to talk openly, I have to tell him that I told everything. If I don't trust you enough to tell you everything, how could I asked him to? I think he has a tremendous amount of information; I don't know what will come of it, but I think the information has to be available to somebody who is trying to make sense of it. The government may know everything anyhow; I don't know. I know they know a whole lot more. I'm sure they are covering with the helicopters, as I mentioned, because those helicopters that fly around out there are sent deliberately. They behave in a manner that draws suspicion. They land out in fields where they shouldn't be. They are trying, almost, to be the guilty party for the mutilations. And the helicopters are not of sufficient size to lift a 2600 pound bull. They just couldn't do it.

Also, you could hardly carry a bull away in a helicopter in a manner so that it wouldn't be seen. If you remember, the mutilated bull on our property was not ours. It did not belong on either side and nobody in the county reported a missing bull. The law officer said that he would ask around, but we talked to him a few months later and no one had reported a missing bull—and it was quite a good one. The law officer said they could have picked it up in Idaho; that was his opinion. No one loses a bull that they have taken that good a care of without raising Cain—it was an expensive animal.

Barbara: It would have had to come through two fence sections to get there. The bones are still out there from both of them; they were never removed.

Jim: There is still the physical evidence too, of the burned spot on the top of the hill, but it is growing back in. I wouldn't go back there to live, because I fear for the family, not for myself. I think it quite likely that something might happen to me. My fear at the ranch is that somebody might get hurt. I understand, as I said, that there have been people who have disappeared. If I had the means, I would have run a full investigation but I don't have the technical expertise, and I don't think I could run an unbiased investigation. I think I got too emotionally involved to keep a clear perspective.

Barbara: I don't think either of us can look at what happened objectively.

Jim: I know that there are people who have the means and technical expertise to run the investigation and I was hoping that somebody, through this group, would be available. That's why I wanted to approach Dr. Hynek. I thought that he might be interested. It's not a onetime shot that can't be followed up; it's going on continuously.

Investigator: Yes, it's a problem to conduct an investigation and to do it discreetly and anonymously.

Jim: And I'm pretty sure you would run into trouble with the Air Force if they knew you were doing it; I have no doubts about it.

Investigator: Do you have any evidence other than the helicopter flights that the military knows what is going on or has been investigating.

Jim: First, I told you I called the military base and they told me to forget it. I first got a radar officer and he was telling me about "unconfirms" on the radar scope, and the next thing I was connected with the colonel who told me to forget it, shut up, and mind my own business. The other time we had Air Force planes overflying us at very low altitudes, and we went down to the base to complain. At the base, I talked to an officer and he said, "Well, what do you think the mutilations are?" I took a very neutral position, saying that it was either the government, some wealthy Satanists, or UFOs. He asked which I personally believed and I told him I believed that it was UFOs. He said that they had their share of trouble with them there at the base too; they had directives on how to deal with them. And then he asked me if I had any trouble with Big Foot. I said that he had been reported in the county. He said they had directives on him too. I said that if he wanted to find him, I could tell him where. He said, "Mister, I don't want to have anything to do with them." So, obviously, there are some directives at the base or I don't think he would give me a line like that.

Barbara: Oh, he did ask you one question. He asked you what the most popular opinion in the county was, concerning the mutilations, and your personal opinion.

Jim: The most popular opinion is the government; the second is UFOs. They figure the government is doing it in helicopters from the military base. Most of the people are wise enough to realize that a group of Satanists, for example, wouldn't have the money necessary to do what's being done, nor could they escape detection. The other thing that indicates that the Air Force knew was the night the interceptors came in low and I got everyone off the ranch. This was the next night after the two National Guard planes went down. There were two squadrons of attack interceptors circling in that area. I got everybody off the ranch because I knew if the disks showed up they would probably come in shooting. It is just supposition that UFOs took them down, but they had two squadrons up there flying around for something. I was with the Air Defense Command and I know when they bring them out, they are nuclear armed and they don't fly them around for games! I counted 26 aircraft and they were circling the whole horizon; they were surveying every thing, supposedly to locate parts of the downed planes. This seemed like a very unlikely thing for high speed interceptors with NORAD markings.

UFOS DEJA VU

The multi phenomena case was appropriately investigated by representatives of the Aerial Phenomena Research Organization headed by Jim and Coral Lorenzen.

Dr. John Derr also investigated the spooky events
at the ranch.

UFOS DEJA VU

APRO is long since defunct and its directors husband and wife Jim and Coral Lorenzen passed away over two decades ago. They authored several paperbacks which can still be found on line. They were particularly interested in encounters with humanoids, as this study indicates.

Dr. Leo Sprinkle, University of Wyoming retired, was the primary investigator on the case.

UFOS DEJA VU

UFOs seem to have a negative affinity toward animals, horses and cows in particular.
Art copyright Carol Ann Rodriguez.

UFOS DEJA VU

Jim and Coral had a very active worldwide membership of several thousand members.

< Seth Breedlove's combination of UFO reports and sightings of bigfoot-type creatures in the Pennsylvania woods is best depicted in his "Invasion on Chestnut Ridge," now streaming on Amazon.

Parody of a popular poster seems to sum up the agenda in just four words.

UFOS DEJA VU

UFOs are dropping down right out of the sky according to those living on
this Colorado ranch. Art copyright Carol Ann Rodriguez.

UFOS DEJA VU

There appears to be a pattern of "over sized" creatures taking place in many cases. We have examined the appearance of "Dogmen" at the various locations.
Art copyright Carol Ann Rodriguez.

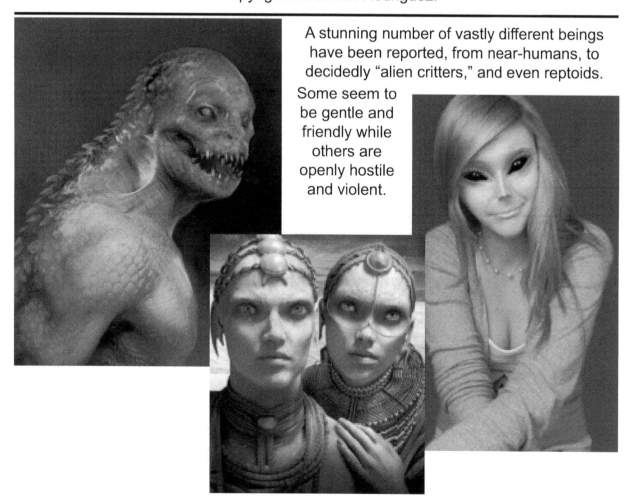

A stunning number of vastly different beings have been reported, from near-humans, to decidedly "alien critters," and even reptoids.

Some seem to be gentle and friendly while others are openly hostile and violent.

UFOS DEJA VU

UFOs of all shapes and sizes were seen haunting the Bradshaw Ranch

SECTION SIX
EERIE DOORWAYS AND WINDOW AREAS

Photo by Ellen Crystal - Pine Bush, NY

UFOS DEJA VU

SECTION SIX

CHAPTER TWENTY-THREE
TOUCHDOWN IN PINE BUSH, YAKAMA'SNOCTURNAL LIGHTS
AND BLACK DOGS ON MOUNT MISERY
By Timothy Green Beckley

CHAPTER TWENTY-FOUR
FIREBALLS, GHOST LIGHTS, AND ALL SORTS OF
ANOMALOUS PHENOMENA
By Timothy Green Beckley

UFOS DEJA VU

TOUCHDOWN IN PINE BUSH, YAKAMA'S NOCTURNAL LIGHTS AND BLACK DOGS ON MOUNT MISERY
By Timothy Green Beckley

If you crave aliens, UFOs or flashing globes or orbs and demand immediate satisfaction, you should consider teleporting on over to Pine Bush. New York. Be sure to pack binoculars, a camcorder with night vision capabilities and a folding chair for an all-night skywatch. But better watch where you park these days cause the cops may not cotton to you hanging around on private property without the owner's permission.

It's hard to say when the next big flap will occur but if you have patience and live along the northeast corridor, chances are your dreams will come true and you will one day catch a glimpse of something truly unusual in this rural community about 90 minutes from Manhattan, or as fast as you can safely drive that convertible in the fresh country air.

This town has become so popular as far as UFO sightings go that the Chamber of Commerce puts on a festival every year, complete with residents attired in alien costumes, characters dressed as the Men In Black, and a John Deere tractor or two just to show you're not in the city any longer.

I suppose I first heard of all the action around Pine Bush back in the early eighties. My friend, Harry Lebelson—now deceased—was working for "Omni" magazine, a spin-off of Bob Guccione's "Penthouse" empire, minus the graphic photos. The publication combined science and science fiction and was brave enough to cover developments in the far stranger arena of the paranormal. Harry was their 'point man' on UFOs, and we shared information that was useful in our parallel investigations. Several times I helped him develop leads, and in exchange he provided me with scoops that might otherwise have passed me by.

Harry had been contacted by a Ms. Ellen Crystal, who claimed to have taken several rolls of "disturbing" photos from a field in Orange County, about 70 miles from the heart of the Big Apple. The photos showed clusters of strange lights and insect-like beings who were only visible on film due to the shortwave radiation the objects and their occupants emitted ... or at least that's how she explained it.

UFOS DEJA VU

It wasn't long before I was invited on a mini-expedition to Pine Bush in search of the elusive UFOs... Elusive to me that is, but not to Ellen and others wise enough to know which areas of this laid back community to hang around, camera in hand.

As it turns out, Ellen had arrived a couple of hours before our van did — mainly because we didn't have good directions and did not know which side roads to take to the prime observation area. When we got to the spot, Ellen was all excited and short of breath. She insisted she had been out filming in the field and had seen a huge cross-shaped object overhead, along with a fleet of intensely glowing orbs. They had paraded across the sky, and she wondered if we had also seen them.

We did not!

Sad as it is, I did return to Pine Bush several times but never did see anything unusual. Many others apparently have. I guess I just wasn't patient enough to camp out in one of the fields before people were being chased by the police under threat of being arrested for trespassing. As usual in a lot of these "window areas," the locals have no interest in welcoming transients. And I have to admit, as it was told to me, things got so out of hand at one point that the local residents had to call the police. Skywatchers were trampling through their yards and frightening their animals and waking everybody for miles around in the middle of the night, as they chased those pesky starcraft.

THE AMAZING ALIEN PHOTOS OF ELLEN CRYSTAL

All you had to do was hand Ellen a camera and she would point and click and its like she was photographing an opening to another dimension – and maybe she was. This brief account of the late photographer's experience is only meant to wet your appetite. To get the complete Skinny on the Minnie, snatch a copy of, "UFO Repeaters: Seeing Is Believing – The Camera Doesn't Lie!"

Truth is, most of the locals were not at all interested in anything that might go bump in the night and thought the UFO spotters more than insane in the membrane.

For several years in the nineties, newspapers and magazines throughout the area were filled with eyewitness testimony. Some of the stories bordered on the truly bizarre despite the fact that they came from such credible sources as Jim Smith, a sergeant at the Woodbourne Correctional Facility, who stated flatly , "I've seen so many of the beings. I know exactly how they move. They're different sizes, different shapes, but when you see them so much, you know they're not of this earth. Not long ago, I saw this figure—about six foot six and dressed all in black— standing beneath a traffic light in Pine Bush. I said to my fiancee. 'What's that woman doing?' My fiancee said, 'Oh, God, I thought I was the only one who saw the thing.'

"When she moved," Smith continued, "it wasn't like she was walking. It wasn't in frames, either, like most of them move in frames; they're someplace and then they're suddenly in another place, like time-lapse photography. But this one moved horizontally. In Pine Bush, you see things you don't expect. I've seen a cat with no head walking across the floor. It had a piece of cardboard where the head should be. A lot of people in Pine Bush tell me they've seen that cat. But not everyone can see the beings. You have to be open to things like that."

One of the things to be noted was the fact that one of the best vantage points to watch for the UFOs, the lights, or whatever they are considered to be, was from the Jewish Cemetery in Pine Bush.

Cemeteries are always good places to see what UFOlogists call "earth lights." Many cemetery sites are picked due to their energy levels, and it is this energy that the forces behind the lights are attracted to or which enable them to become viewable in our spectrum.

Another spot is a Native-American burial ground.

The UFO spotters mostly don't realize where they are standing, because obviously there are no markers to designate the graves of the Indians. I had a girlfriend whose relatives lived in the area and they said they had seen weird lights in the fields since they were kids But they wanted nothing to do with the "crazy" UFO people and would never admit to them that they had seen anything strange. Before visiting Pine Bush, I suggest you stop by PineBush.com for directions, and if you get to town before nightfall, stop in at Butch's Barber Shop and ask Butch what's going on. Last time we were there, the walls of the shop were covered with framed clippings and photos of the phenomenal light patterns. And check to see if my name is still on a plaque somewhere in the shop, requesting a bit of feedback. The plaque may be in the bathroom – no shit!

Ellen Crystal is deceased – seems like a lot of good UFOlogists die young. For a long time before moving out of the area, Ellen had been assisted by Bruce Cornet who later worked for Bob Bigalow in association with the Skinwalker Ranch. Dr. Cornet came to explore the back roads of Pine Bush so thoroughly, that it got to the point where he could almost predict when a UFO would be coming overhead. He video taped the arrival of these craft as they came through a portal that must exist in Pine Bush, though not noticeable to the human eye. To avoid detection, some of the craft have learned to "come through," so as not to be recognized for what they truly are, disguising themselves as jet aircraft no less. This is the morphing phenomena we have seen so often so you can't identify the face behind the mask.

If you want to find out what Bruce is up to, readers are invited to go to an interview we did with Bruce on our Exploring The Bizarre show originally broadcast over KCORradio and archived under the episode entitled "Invasion Of The Saucer Men – Pine Bush, NY" on YouTube at

www.youtube.com/watch?v=i_H_qsaFfHc&t=4036s

UFOS DEJA VU

* * * * *

YAKAMA'S NOCTURNAL LIGHTS

If there is an emerging pattern to indicate from where UFOs – sometimes identified as "earthlights" – can best be seen, and from whence strange energies can best be felt, and where windows and doorways to other dimensions open and close most frequently, I would have to say that North American Indian Reservations would have to be the A-number one spot for encountering these anomalies.

On my frequent trips to the Southwest across various Indian reservations, there was always some chatter among tribes-people regarding what we would call UFOs but for which they have different tribal names. The Native-Americans who live on Arizona's Second Mesa have all sorts of legends about the sky gods who came down and helped them with the planting and gave them medical assistance which kept them alive during droughts and plagues. Back in the seventies, groups made up of Hopi as well as sympathetic outsiders would gather on the Mesa to await the return of the Kachinas. If you have ever been to a powwow, you will have noticed strange figures and small statues for sale that look truly alien. These are the Kachinas, who are representative of the long-departed star people.

If you are ever vacationing near Toppenish, Washington, (better get a good atlas to find it), try to make it up to the Yakama Reservation in the Hembre Mountains and drive out to the Zillah-Toppenish Road. There is a viewing spot near the cemetery from which nocturnal lights can be viewed.

I am not sure who you can check with—you might try UFOArea.com—but it's said that some seasons are better than others when it comes to seeing the lights on the reservation. David W. Akers seems to be the most active investigator in the area. They did have an observation post there for a long time. I have no idea if it's still standing or not. Sometimes the lights stand still. Other times they are fast moving. In a report filed with the Center For UFO Studies, researcher Akers noted a number of patterns: (1) A tendency for activity to occur in groups or 'waves.' (2) Type of activity is not limited to nocturnal lights, but includes other types of UFO phenomena, and (3) Activity that is characterized by an evasiveness which exceeds normal expectations of chance."

The list of Indian reservations on which active doorways may exist include, but are not limited to:

Ute Indian Reservation in northeastern Utah. Strange lights have followed travelers down isolated roads. It's pretty eerie after dark. Not many street lamps. This would be the general location of Skinwalker Ranch.

Papagos Indian Reservation and the area around the Superstition Mountains, both located in Arizona. Travelers visiting the Superstition Mountains are warned that miners and tourists have mysteriously disappeared while on foot looking for the Lost Dutchman Mine. The massive wave of UFOs over Phoenix down to Flagstaff took place not far from here. And no trip to experience the strange would be complete without a vision quest in Sedona, Arizona, especially at Bell Rock on a full moon. Go back to the chapter on the Bradshaw Ranch to see what I am talking about.

Quite a bit has been written about all the strange activity that has been going on for decades near Dulce, New Mexico. "It's a window to another dimension already," Beckley explains, "but it has often been associated with the dark side of the unexplained. This would include animal mutilations, psychological warfare on the part of the UFOs and supposed underground bases." Beckley says he doesn't think the Native Americans living on the Jicarilla Apache Reservation appreciate the beer can tossing tourists who started flocking to the area disturbing their tranquility after word of the phenomenon leaked out.

Another Indian reservation where the macabre is not out of place would include the Wind River Indian Reservation in Wyoming. "A lot more investigation needs to be done on the phenomenon of earthlights and such in these locations. For example, are these focal points on grid or ley lines? To be sure, the natives are very tuned in to Mother Nature—having lived under the clouds and stars—so naturally they would be more likely to be able to understand her nuances."

The following are detailed reports on window areas that seem to lead to other dimensions and which deal with a variety of anomalous phenomena reported to have taken place in locations considered sacred to Native Americans. These locations should certainly be placed in the HIGH STRANGENESS category.

* * * * *

THE MIB AND BLACK DOGS OF MOUNT MISERY

One of those locations is known as Mount Misery, on Long Island, New York.

According to a newspaper called "The Long Island Press," the area has been notorious for strange events for over one hundred years.

"Many different tales of horror and spirits dwell around this place," investigative journalist Christopher Twarowski writes, " from a hospital that burned down in the 1800s to a group of kids hanging themselves under an overpass near there, to ghostly apparitions of 'The Lady In White,' named Mary, who was hit and killed by a car on the road. The most believable concerns the escaped mental patient who used his prosthetic hook to kill high school kids who were using the area as a lover's lane. "I drove the length of the road," Twarowski adds, "but I didn't see anything but million-dollar homes. Definitely spooky'."

But there are other witnesses in the Mount Misery area who have seen more than opulent houses. How does a Man-In-Black strike your fancy?

"One of the most well-documented reports," says a website called "Long Island Oddities," "came from a woman living near the peak of Mount Misery. She had four gentlemen come to her door in the rain. They had walked up a muddy hill but their shoes were spotless. They had strong, high cheekbones and seemed to be of Indian heritage. They spoke politely but claimed that her property was theirs and that they would get it back."

"There have been other reports of Mount Misery natives seeing strangers in black suits wandering around," he goes on, "though I have been unable to gain any more specific info." Many of these events took place many years ago, during UFO sightings flaps."

One online writer, calling herself "Nikki aka Witchlight" has written about strange happenings in Mount Misery by way of nearby Sweet Hollow Road.

"The road has been surrounded by stories for generations," she writes. "There are tales of 'The Black Dog of Misery,' whose name might be a hint at the closely located Mount Misery . It is said to appear as a black Labrador retriever type dog. It is evil, according to stories I've been told, but rare. To see the dog is a harbinger of impending death. There are even tales of a phantom car that drives down the road at breakneck speeds, careening off the edge of the road into a flat, marshy section of land. When people get out to investigate, they find nothing, even though they thought they witnessed the whole thing."

And if you feel like taking a touristy trek into the unknown, Nikki provides some directions on driving there.

"The more I asked about Sweet Hollow," Nikki continues, "the more stories I got. It's rare to find a native Islander who has never heard of some story pertaining to the road. It can be found off of Jericho Turnpike in Huntington. It also runs past the Northern State Parkway, Old County Road, Walt Whitman Road, and runs into Pinelawn Road at Route 110. Therefore, it should be easy to find from almost any major roadway on the Island. Watch your rearview mirror carefully, and keep telling yourself that the man you see walking along the shoulder is real, or at least just your mind playing tricks on you."

UFOS DEJA VU

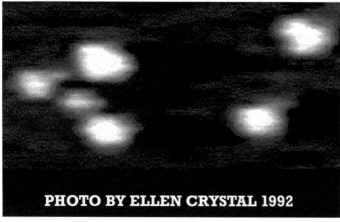

PHOTO BY ELLEN CRYSTAL 1992

Bruce Cornet writes — The "Westchester Boomerang" was seen during the 1980s over the Hudson Valley, and again videotaped in 1992 just west of Chester, NY, by the late Dr. Ellen Crystal, with whom I worked."

Dr. Cornet arrives in Pine Bush to investigate an important UFO incident. Full report on his findings can be found at Eleven Years After The Pine Bush Flap. http://www.sunstar-solutions.com/AOP/Appendix_I/APPENDIX_I.htm

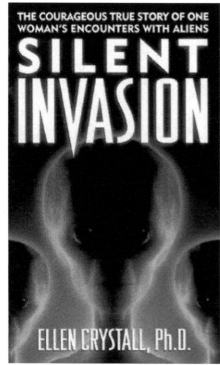

< Ellen Crystal's book that detailed the hundreds of reports from the Hudson Valley area.

Ellen Crystal interviewed by Japanese TV crew in Pine Bush, NY.

UFOS DEJA VU

Pine Bush Jewish Cemetery where many UFOs have been observed.

Photo taken April 28, 1993 by Dr. Bruce Cornet at Pine Bush Vortex.

Ellen Crystal's remarkable photo of insect-like UFO beings in open hatchway of UFO in Pine Bush from "Silent Invasion."

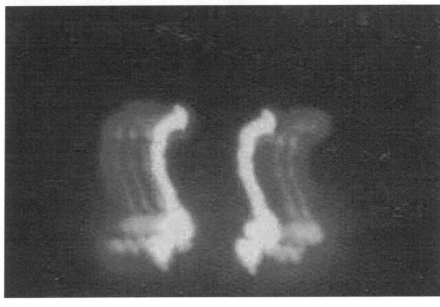

UFOS DEJA VU

Tim Beckley throttles neck of alien space captain at Pine Bush UFO Festival.
That will teach him!

Dare you travel along Mount Misery Road — these crows look like they might welcome you! — www.gothichorrorstories.com/

UFOS DEJA VU

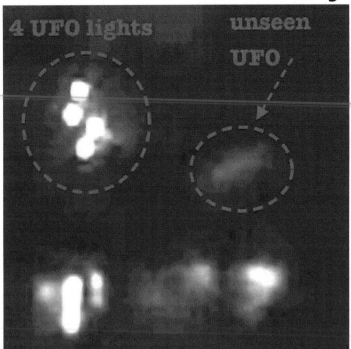

4 UFO lights

unseen UFO

The mysterious Yakama, Washington lights on Native American Reservation.

WBAB talk show host Jaye Paro's article on Mount Misery in Beyond magazine. courtesy JohnKeel.com

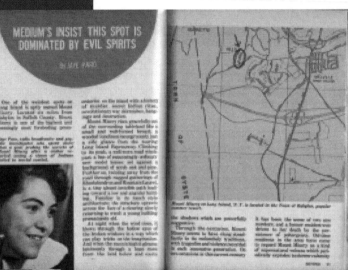

MEDIUM'S INSIST THIS SPOT IS DOMINATED BY EVIL SPIRITS

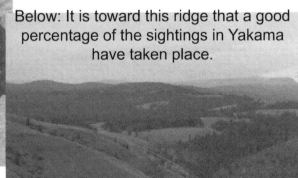

Below: It is toward this ridge that a good percentage of the sightings in Yakama have taken place.

YOU ARE NOW ENTERING
YAKAMA INDIAN RESERVATION

At one time a UFO viewing tower had been constructed to watch for the Yakama UFO. The Yakama Indian reservation is the gateway to where the best sightings have transpired.

UFOS DEJA VU

FIREBALLS, GHOST LIGHTS, AND ALL SORTS OF
ANOMALOUS PHENOMENA
By Timothy Green Beckley

Step up to your local and global gateway, and let's get the hell out of here. Dimensions unknown!

If you consider yourself a world traveler and don't mind long flights, pack your bags and head to the Mekong River in Thailand. Go there during the first waning moon any October (usually the 22- 23) and you can take part in an officially sanctioned skywatch attended by tens of thousands of devoted Buddhists.

Around eight in the evening, get as close as you can to the river bank in the town of Phon Phisal and arch your eyes skyward and gawk in awe as the mysterious Naga Fireballs shoot up from beneath the water then disappear before they get to the other side of the river in what would be Laos.

The Bang Fai Phaya Nark, as the monks call them, are totally silent and odorless, and despite skeptical beliefs that they may either be methane gas from decaying plants or tracer bullets fired from the jungle, no one is about to argue with the throng of believers who gather for the carnival-like festivities which are part of the celebration.

It's a ritual that has gone on for hundreds of years, and each autumn the crowd of the curious continues to grow as word of the light works spreads beyond the country's boundaries.

The best observation point is near a 450-year-old temple that can easily be reached on foot. One observer stated that they saw the fireballs "come up from the river, and also from the canals and dams" and that they were convinced of their supernatural legitimacy because not only had they seen them for half a century, but their mother and father also witnessed the unexplainable fireworks display of Mother Nature.

One travel brochure phrases the proceedings this way:

"On the last day of the Buddhist Lent, the people of each village located near the river at Amphoe Phon Phisai float their fireboats as a mark of respect and tribute to the Lord Buddha. They then place torches made of a substance of insect waste collected from trees on their 20-30 meter long bamboo boats to illuminate them. The people all gather together at the harbor in front of Wat Phadung Suk Nue. Once everything is ready, they light up their torches and launch their boats.

"In order to make their boats look more spectacular, some team members fire small Bang Fai (rockets) and the circular Bang Fai known as 'talai' into the sky. It is then that a curious and so far unexplainable phenomenon occurs – some Bang Fai were fired as if from under the water, as if the Paya Nagas wanted to participate in the event. This has gone on for hundreds of years and the phenomenon has occurred every year since."

The traveler's guide concludes that, "This festival has been passed down from generation to generation, from father to son, as witness to these unnatural phenomena. It is as mysterious as it is spectacular, and this is the only place in the world where such a thing happens."

Almost makes you want to leave on the next red eye!

A DIFFERENT KIND OF BEAST

No shapeshifter or skinwalker onboard here, but we do have a paranormal tie in – what else did you come to expect?

What makes this even more of an enigma is the appearance along with the fireballs of a sea creature that the locals liken to the Loch Ness Monster (Hey, Nessie even has friends halfway across the world it would seem), except they boast that they have captured the Paya Nagas – not only on video, but in the flesh. A travel report we have read says, "An American serviceman caught a King Nagas. Of the men who hoisted the fish, only eight or nine survive to this day. They had a reunion in Udom Thaini Province three years ago. Five died in the Laos war, and many of the others have died in the years since the creature was captured, at least three of them horribly."

I guess we are to believe it is sort of bad luck to have handled the Paya Naga, kind of like a King Tut's curse on water.

The same report states that, "The beast has the head of a dragon and a body 7.80 meters long. It was sent to America for investigation, but unfortunately did not survive. Nevertheless, it was authenticated simply by its capture by the American serviceman."

The experts at an American piscatorial society examined it meticulously and found: (1) That this was a unique creature, not seen or captured before. (2) That it was a creature of the river and not from the sea. (3) That it had green blood.

There is a "highly suspicious" photo of the supposed creature, but we cannot reproduce it since it holds a 2002 copyright even though it is said to have been taken back in 1968. Everyone seems to take the existence of the creature in a good-natured way, regardless of the roots of its subterranean origins.

One thing is for sure – don't go there expecting to badmouth the Bang Fai, as you might end up in serious trouble.

Australian journalist Gary Walsh notes that after a Thai TV station aired a segment claiming that the fireballs were tracer rounds fired from AK-47s by Laotian soldiers, all hell broke loose with those who felt their religious beliefs had been slandered.

"Hundreds of funeral wreaths (to curse the television station) were floated on the Mekong by protesters at Nong Khai, the capital of the province where the fireballs are seen, and demands were made that the provincial government sue the TV station for a billion baht – about 35.5 million in Australian dollars."

Writing for FairfaxDigitaL com, Gary Walsh says the episode became "a full-blown international incident. The Laotian ambassador to Thailand expressed his country's dismay at the claim that some of its citizens had manufactured the fireballs."

The papers in Bangkok were likewise "full of angry letters, and TV news bulletins covered the furor with lip-smacking enthusiasm."

The TV station ultimately had to issue an apology and five Laotian villagers who had helped the documentary team were reportedly given 12 years in jail for bringing foreigners into the country without permission and endangering Laos' security. One of the villagers, a militia member, fired the weapon that "proved" the Naga was a hoax.

We wonder if the day will come when members of the media in the U.S. are taken to task for being critical of UFOs or paranormal happenings involving credible witnesses?

* * * * *

PLACES TO GO AND WEIRD THINGS TO SEE

As you ponder the mysteries associated with earthlights, spook lights, ghost lights, or whatever you prefer to call them, it may occur to you to actually visit some of the locations where the mystery lights are frequently seen – a kind of tourism of the paranormal where you pack a cooler of soft drinks and head for the highway with the kids in the back seat and the radio blaring the latest from the Top 40.

What follows is a listing of locales that feature regular appearances of spook lights. Details not covered here can be found on sites such as "About.com," in its section on the paranormal.

Big Thicket Ghost Light in Bragg, Texas: This light can be found along Bragg Creek, near the old ghost town of Bragg in Eastern Texas. Viewed on a dirt road that leads into swampland, this spook light is accompanied by a famous legend regarding a railway brakeman who was accidentally beheaded by a passing train and has been searching for his head by lantern ever since. The light is described as pumpkin-colored and starting as a pinpoint of light among the trees that in-

UFOS DEJA VU

creases to the brightness of a flashlight before it fades away.

The Gurdon Light in Gurdon, Arkansas: This ghost light has been witnessed by hundreds of people who live near or travel in the area of this small Arkansas town, about 75 miles south of Little Rock. One has to hike about two and a half miles off the road to the spot where the light is seen, which of course eliminates headlights as a possible cause. The light is usually seen as white, blue or orange, and seems to have some kind of distinctive border around it.

The Hebron Light in Hebron, Maryland, has been seen for decades. It is reported to have appeared as a ten-inch ball of light that led two officers of the Maryland State Police on a merry chase in their patrol car down a dark road. There have been sporadic sightings of the light before and since, though some claim that the light has been inactive since the mid-1960s.

Of all the fifty states, New Jersey has a deserved reputation for being truly strange. It even has its own periodical devoted to the unexplained and the peculiar. "Weird New Jersey" has done more than its share of stories on spook lights (like the ones in cemeteries), but the most persistent earthlight phenomenon is associated with the town of Flanders, where teenage couples have hung out for decades to catch a glimpse of Hookerman. Seen along a trail where a stretch of railway once existed, the lights appear singularly, in pairs or as a loving threesome. Most often they are observed bobbing below treetop level. The steel spikes and cross rails are long gone, and the trail is now under the supervision of the parks department. This does not, however, prevent ghost light watchers from setting up shop with their binoculars and cameras. In the last couple of years, the lights were caught on video by the crew of "Hometown Tales," a website and TV program devoted to urban legends of the Garden State. Reports have it that the lights are caused by the lantern of a railroad worker killed on the spot, but we've heard that "off kilter" fallacy a dozen times before, and besides, there are no historical records indicating such an accident ever occurred at or near the location.

The Marfa Lights, in Marfa, Texas: Located in western Texas, nine miles east of Marfa, these lights are arguably the most famous the U.S. has to offer. The lights have been seen for more than a hundred years. According to an official tourist brochure, issued by the State of Texas, the first recorded sighting was made by a rancher named Robert Ellison in 1893. Today they can be seen at night by observers who park along Highway 90. The lights are said to change in color and intensity and usually move about. The town of Marfa hosts an annual Marfa Lights Festival every September, and has announced a new Marfa Lights Viewing Center as well.

The Min-Min Lights, in Boulia, Queensland, Australia: If you're in the mood for some international tourism, check out the Min-Min Lights in Australia. Named Min-Min after a hotel from which the lights were first observed by white settlers, these

244

UFOS DEJA VU

ghost lights have been seen for many years on Aboriginal land. The lights seem to "follow" witnesses as they drive in their cars. They vary in color and are sometimes oval in shape and exhibit erratic movements. Some Aborigines view them as ancestors, gods or demons.

The Ontario Ghost Road, in Scugog Township, Ontario, Canada: Mysterious white and red lights have been observed by several witnesses who brave the "Ghost Road," also called the Mississauga Trail on Scugog Island. The legend behind this ghost light concerns a motorcycle rider who lost control of his bike while speeding down the road in the 1950s and fell and hit his head on a rock. Ever since the ghost lights of his motorcycle have been seen to cross the field, make a turn and then disappear.

Here is a job for those independent and MUFON state and regional directors reading this, to get out there and scope out some of these window areas to see if they are actual portals or gateways to other dimensions, or if observers are seeing more or less just natural phenomena. There seem to be multiple phenomena taking place here and about, but we need to have boots on the ground to determine the extent of what's going on.

Get back to us when you can.

Some say a legendary sea serpent is responsible for the lights along the river.

UFOS DEJA VU

Mekong River phenomenon attracts thousands
on a specific date to witness the other-worldly transmissions.
Researchers are conflicted as to what causes the lights to appear.

Bragg, Texas big thicket light — the mystery is on going.

UFOS DEJA VU

Gurdon Light—Gurdon, Arkansas. Better get off the road.

Mae Clair says: "A light phenomena of the Queensland region of Australia, the Min Min is a large flickering disc of luminescent light that appears at night, hovering about three feet above the ground. Named after a small settlement in the Outback, the Min Min made its first appearance in 1918."

Min Min Light road sign, Queensland, Australia.

The media has given the Min Min Lights a lot of attention.

UFOS DEJA VU

The Marfa Lights—you are welcome to come and see them. Film crews from around the world have come to Texas to capture Marfa's Mystery Lights.

The Marfa Lights at Marfa, Texas have been observed for decades and are still unexplained.

High strangeness doesn't get any spookier than this!

SECTION SEVEN
TRACKING THE MYSTERY LIGHTS

UFOS DEJA VU

SECTION SEVEN

CHAPTER TWENTY-FIVE
DR. J. ALLEN HYNEK AND NORWAY'S MYSTERIOUS LIGHTS
By Timothy Green Beckley

CHAPTER TWENTY-SIX
TRIPPING THE LIGHTS FANTASTIC
By Erica Lukes
Introductory comments by Tim R. Swartz, KCORradio.com

UFOS DEJA VU

DR. J. ALLEN HYNEK AND NORWAY'S MYSTERIOUS LIGHTS
By Timothy Green Beckley

Let's rule out right away the northern lights – the aurora borealis – let's just take that explanation off the table. This is NOT a "glowing sky" phenomena. This is something vastly different. Something totally unique. Something completely out of this world!

Strangely enough, indeed, that a small valley in central Norway named Hessdalen may be among the world's most popular places to see strange "lights" in the sky – and on and near the ground for that matter – on a regular basis, going back a minimum of four decades and attracting the attention of researchers the world over who want to add a notch in the form of an actual sighting onto their UFOlogical belt.

The late, legendary UFO investigator Dr. J. Allen Hynek journeyed to Norway just before his death to observe the enigma for himself. As with so many other kinds of paranormal developments, the good astronomer didn't undergo a transformational experience, as he was only in Hessdalen a couple of days and the objects were playing hide and seek with him.

The objects, however, do seem to appear so dependably that one is almost guaranteed to see something, especially during the long Scandinavian winter nights, if they "camp out" inside where the temperature is not below freezing. When DR. Hynek visited, it averaged about 36 degrees below zero. The Hessdalen Valley is 12 kilometers long and a scant 150 people actually live there. But in December 1981, the relative peace and calm of life in Hessdalen was suddenly and irrevocably changed forever.

According to a website "Project Hessdalen— "In December 1981, unknown lights suddenly started to show up. The lights would stand still for more than an hour, they would move around slowly, and sometimes they would simply stop." The lights were also capable of traveling at high speeds. A speed of 8500 meters per second was tracked by radar on one occasion.

'The lights could be anywhere," Project Hesssdalen professes. "Sometimes they were reported to be just above the roof of the houses, or just above the ground.

UFOS DEJA VU

Sometimes they could be high up in the air. Mostly the lights were reported to be below the tops of the mountains nearby. No one could give an explanation for these lights."

There were many interesting variations on what the lights were reported to look like.

"The lights were reported to have different forms," it's maintained. "This variation showed up on the photos taken of the lights. A light could have the form of a bullet, with the sharp end down. It could be as round as a football. It could look like a Christmas tree turned upside down. The colors were mostly white or yellow-white. Sometimes it could have a small red light, usually on the top or the bottom of a white light. As Nils Magne Ofstad, a researcher on the lights and brilliant author astutely puts it, "Residents don't only see lights, eventually the brave people living there told about Christmas shapes like saucers and eggs or eggs and triangles, these being so close that they could throw rocks at them."

"A few times," the report from Project Hessdalen continues on, "it seemed to have all the colors combined: red, green, blue and yellow at the same time. Sometimes on the photos you can see a distinctly blue light."

The lights do appear as stated quite frequently, but there are some seasonal variations

"The lights were occurring several times a day, but mostly during the evening and the nighttime. At the most, lights could be seen about four times a day. There were more lights in the wintertime. In the summertime, lights were seldom seen at all. One reason for that might be that there is daylight almost the whole night in summertime in Hessdalen."

"What is this phenomenon?" the Project asks. "What do we know about it? We have not found out what the phenomenon is. That could hardly be expected, either. But we know that the phenomenon, whatever it is, can be measured."

Which of course seems to prove that something real is going on. Whether it is some unknown kind of natural phenomenon, UFOs, or even the same kind of ghostly phenomenon often thought to explain similar lights here in America, the lights of Hessdalen are something genuine and palpable. Hynek had to leave the Hessdalen lights behind as still another unknown. But if you're willing to brave central Norway in the dead of winter, you may be able to see some of the most dependable spook lights in the world. And if you decide that you can't make the trip, you can at least take a look through the gateway as presented in the pages of this book, eh?

DAYLIGHT SIGHTINGS

Nils Magne Ofstad is on top of the action. His "Hessdalen Lights: What's Happening In The Norwegian Mountains?" has only been out a couple of months and it is certain to stir attention in the international UFO community – or at least it should

UFOS DEJA VU

as it offers overwhelming evidence that the mountain village is an incredible portal deserving continuous monitoring. This is done through the use of a wide variety of cameras and computer systems which have been installed in the Blue Box UFO Laboratory on the spot where the objects can be observed coming and going over and in the valley on a regular basis. Nils, who is a senior lecturer, has had a long interest in the UFO phenomenon, and in particular Norway's own hotspot, Hesssdalen, being a two hour drive from his home in Trondheim. He is preparing a sequel to his first book.

His orderly research covers many areas of research convincingly, in particular the advanced instrumentation work being done on what has been dubbed the "Project Hesssdalen Campground." For example:

** – How scientists involved in the project conclude that a significant amount of data indicates illuminated solid objects.

**— That Hessdalen residents insist that what they see are physical crafts - not "just" lights.

**— how some of the scientists have experienced close encounters as well - including 'daylight discs'.

** —That scientists with basically a geophysical approach have "converted" in their thinking as to the origin of the phenomenon - and openly take the extra-terrestrial hypothesis (ETH) into consideration.

**—The alleged UFO landing - and the sighting of a humanoid on the ground - at mount Rognefjell in Hessdalen.

**— Examples of lost time (possible abductions?) in connection with UFO encounters in Hessdalen. And,

**—Speculation about what Hessdalen might represent on a broader scale.

Nils has gathered many sightings that are part of his book, but here are a couple taken directly from his blog Hessdalenblogspot which will indicate the caliber of the reports:

LARZ LILLEOLD

This evening Lars had an errand outside his house. The time was approximately 19.30. - "Over the outhouse I saw something big, bright red. It must have been at least 30-40 meters long. As a loaf of bread it stood there, completely still. I expected to hear sounds. I listened, but heard only the murmur of the forest. Then 'it' moved. The light was blinding. It looked as if "it" had a red, hairy like smoke that it surrounded itself with. It moved in a northerly direction up the valley." He shouted at his wife who saw the lights before "it" disappeared. - "Since it was a craze for me, I had to see this again. And I have done so many times later. Time and again I asked myself what this might be. I'm surprised that the authorities do not take this more seriously. I'm glad that others see, and want as many people as possible to do so."

UFOS DEJA VU

JON ASPÅS

A man who has seen UFOs both day and night, is Jon Aspås. In late February, he called his neighbor, Martin Aspås, who lives a short distance further up the valley. The time was quarter past nine in the morning. Martin sat and ate breakfast and looked out the kitchen window when he saw an object in the air just to the south, in the direction of the houses to Jon Aspås' property. "Do you see anything over mount Rogne? Is it the same object that we have seen in the evening?" Jon ran outside the house. There it was, shiny and metallic. It was oblong and somewhat flattened below. It looked like it had a raised platform at the top, says Jon. It went south at an altitude that could not be more than a few hundred meters.

The following morning he saw the UFO again. Then it was also observed by others near Røros. At this second daylight observation, they could study the object for 42 minutes, as long as it stood quietly in the valley between Ålen and Hessdalen. Then it went straight east before it sat up a tremendous speed and disappeared out of sight. Jon saw the UFO once again in March. It was passing the valley in daylight, but very fast. Once it went four times between the valley and mount Finnsåhøgda, back and forth.

BERIT MARY KIERRENGVOLD

Berit Mary Kjerrengvold lives in Hessdalen. Tuesday before Easter she was on her way from Ålen towards Røros in her car. With her she had an acquaintance, Anita Trygstad from Røros. Approximately at 13.00 they were a mile south of Ålen, on a slight incline. On the right side, they discovered something dark that was not moving over the treetops. The object appeared elongated and was sort of standing at an angle. It was metallic silver and quite large. The sun was up, and should have reflected the object's metallic color. It did not, it looked more as if the metal had its own "glow." They did not stop the car and watched the object for approximately 20 to 25 seconds. In October, she and her husband saw a large object with two bright lights in front and a red at the rear.

FOOTPRINTS IN THE SNOW

An excerpt from the book "Hessdalen Lights" (2019): "…according to the old guest book on Project Hessdalen's website, a Oslo couple in their twenties allegedly saw a UFO landing here when they came driving the Hessdalen main road a winter evening in the late 90s. The craft was supposed to have been as big as the main house on the farm and flashed in all colors like a Christmas tree. The couple made their way to the farm the next day, and apparently found imprints in the snow, - both after the UFO and some kind of "footprints" next to them…" In the observers' own words: "- The next day we headed up to Hessdalen to look more closely at the place we saw the UFO. About 200m from the road, we found tracks in the snow. There were three 'landing legs' (with footpads?) that had melted down in the snow, - right down to the ground so that the grass could be seen. In a cir-

UFOS DEJA VU

cumference of approximately 50m there were some strange footprints. There had not been other people there, then we would have seen (normal) footprints. What we saw the night before must have come from above."

The story of the lights is impressive and gets stranger and more exciting as we go along on of Hesssdalen the trip to Norway and a visit to the village with Erica Lukes.

SUGGESTED READING

HESSDALEN LIGHTS: WHAT'S HAPPENING IN THE NORWEGIAN MOUNTAINS?

http://ufohessdalen.blogspot.com/

While in Hessdalen, Dr. J Allen Hynek had to endure temperatures of 36 degrees below zero.

Researcher and author Nils Magne Ofstad keeps tabs on the mysterious lights.

UFOS DEJA VU

Finally, a book has been published that tells the world outside the valley what is going on in Hessdalen.

Equipment monitors aerial activity in the valley regularly.

Nils and Dr. Erling Strand in front of the Hessdalen Lab.

UFOS DEJA VU

TRIPPING THE LIGHTS FANTASTIC
By Erica Lukes
Introductory comments by Tim R. Swartz, KCORradio.com

PUBLISHER'S NOTE: Erica is among the most energetic and enthusiastic "new breed" of UFOlogists. Even if you don't know the name, you would recognize her immediately from her various appearances on the Travel Network, and the Discovery channel. She is a recognized "go to" consultant for many of the programs that are airing this season. And well she should be, as she is most knowledgeable, well read and widely traveled. She has "laid low" on Skinwalker Ranch (see previous chapter) and shared her knowledge with listeners to her Friday evening "UFO Declassified" podcast over KCORradio.com, the same Las Vegas station that airs "Exploring the Bizarre," with Tina at the helm as producer and engineer.

On a recent ETB episode, my cohost, Tim Swartz, wanted to find out the nitty gritty about how she ended up in Hessdalen, Norway to try and view the phenomenal light brigade that the village is known for, to which Lukes (who teaches Pilates by day in her Salt Lake City studio) had this to say: "I was invited over by Professor Erling Strand, who is with Oswalk College University – and he is, in my opinion, probably the most important and groundbreaking researcher that we will ever know in this topic and one of the most humble human beings I have ever met. And I was invited over there to visit their Science Camp. Twice a year they bring students from the College of Engineering. They go there to the town and they climb to different places in the valley and this is under very harsh conditions. We're talking in the middle of the winter. They climb up to the tops of these mountains where they find the best observation points. They set up their equipment – and they have plenty of it. They have funding from different universities in Europe and they are gathering all sorts of information. Thirty five years ago when Dr. Erling and his colleagues first decided that they wanted to do this, they had to look at a specific set of data. They couldn't just go there – and even though they were getting reports of A, B, C and D, they had to decide what they could do to really get started and funded in a very scientific manner so that they could look at this anomalous light phenomena to determine its origins and the circumstances

UFOS DEJA VU

under which the lights appear most frequently."

Erica told Tim Swartz and our "Exploring the Bizarre" audience that they were anxious to determine if there was an intelligence behind the lights. "Dr. Erling identified four different types of light phenomena that behave intelligently or as if they're intelligently controlled. So this is big. And nobody in America knows about it because we're too busy talking about Area 51 and all those things that at the end of the day are probably just fluff, and not contributing anything to the actual study of what's really taking place."

One of the most fascinating individuals the media maven was introduced to is a Jon Arvid, a gentleman who has lived in the valley his whole life. "There's a really great documentary called 'The Portal," which I would recommend everyone watch. They were interviewing Jon about these balls of light that were appearing outside their home. He talked about the fact that they were seeing scoops of soil that were lifted out in a cookie-cutter shape. They were seeing structured craft (not just lights in the sky!). He mentioned that there was a person who saw a small entity. So they're seeing again, the same type of things they're seeing at Skinwalker ranch. It was really a remarkable experience to sit across from Jon in this tiny valley and listen to his experiences as an 80 year old and listen to how that affected him. And how the community – just like Skinwalker, just the Basin-how the community was in a way manipulated by the media and also really did not feel like they had a place to turn. So that's why it was remarkable to see Professor. Strand get in there and understand that he could use science and the scientific method and give these people a sense of security."

To say the least we were flabbergasted by what Erica Lukes had to say, but we quickly ran out of broadcast time. So when we mentioned that we were preparing UFO Deja vu and were going to include a section on the Hessdalen phenomena we did have to do much in the way of arm twisting to get Ms. Lukes to extend our conversation in the form of a written interview. What she had to offer is presented here exclusively for our readers.

Rock on. . .

* * * * *

In September of 2017, I spent two weeks in Hessdalen, Norway, observing Project Hessdalen Science Camp. Students from the engineering college at Otsfold University and world-renowned scientists meet twice a year in the picturesque valley located in central Norway. Their goal is to collect data on "The Hessdalen Phenomena."

I had the opportunity to sit down with Professor Erling Strand, who spearheaded the most critical research project in the world with regards to anomalous light phenomena. His words have left a profound impact on my work. Here is a portion of the interview.

UFOS DEJA VU

LUKES: So, tell me about, I know you've told me the story a million times, but how this all came into being, how 35 years later you're still here?

STRAND: Well, it started already back in the Eighties. In 1981 there was a considerable number of sightings up here in this small Norwegian valley 50 kilometers long, and the people saw mostly a light outside and were wondering what it was. Newspapers wrote about it, and several people came up here. At the beginning of the 80s, this lasted a long time with this high intensity.

Well, from November 1981 the light continued until late 1984, so many people went up there, and I was one of those, together with some friends. We wanted to see.... for myself...for ourselves. And we did see it, and it made a big impression on me because it was a huge light. That was back in 1982. No one did do anything, just talked about it. And people were asking why that university doesn't do something, why doesn't that research institute do something. Everyone talked about that, and we also did. My friends and I said, "Well, why don't we do something. Yes, we can do something, why not?"

Three decades later, Stand and his team have inspired many young researchers. On my journey, I accompanied Dr. Strand to three different locations where students and scientists collect data. We visited the "blue box," a large container that's the heart of the project. I met with different European scientists, including Dr. Jader Monari and Dr. Stellio Montbuliani, from the National Institute of Astrophysics in Bologna, Italy.

In the evening, I accompanied Strand to a remote location where they've had a significant number of observations over the past 35 years. We set up our cameras in the hopes that we could catch the Hessdalen Phenomena and battled the wind and frigid temperatures.

I also interviewed Jon Arvid, a native of the Hessdalen, who had witnessed the phenomena for decades. Mr. Arvid has lived in the Hessdalen Valley his entire life and graciously shared his experiences with the Hessdalen Phenomena.

(Special thanks to Tomas Dahl for translating this interview)

LUKES: When did you start to interact with the phenomena?

ARVID: Hmm, now I have to start counting. It was February 82. I'm born in 36 so you can start counting ha-ha. Forty-six years, that's right.

LUKES: What did you see?

ARVID: "Ha-ha, yes. That's a long story because that's when we saw 4 of them with Åge (Moe) and Bjarne (LIllevold). We saw 4 of them at the same time. We were on these mountains further in (points towards west). There we saw three which were on the ground or looked to be on the ground. Completely calm. Not moving. Then the one in the middle started to move and switched place with one of them. After they stood there for a while, one of them went straight up and stood up there for a while, about 100 meters up in the air. All of sudden the light became very strong and shot right up. Then another light came down and took its place. The light that went straight up, flattened out and came right above us where we stood. We were 3 persons who drove snowmobiles in the mountains. And when it was right over us, the Moon was about to come up. Then the moonlight reflected the object which took off."

LUKES: Where you frightened?

ARVID: No no, I wasn't scared. But I didn't sleep too well when I got back home. Not because I was scared but I was wondering what in the world had we seen! Because the 3 lights we saw on the ground, we watched them for almost 45 minutes. One of them had luminous particles around it, similar to headlights that people wear today. We were on a distance about 4 kilometers and the lights were in the next valley from us. But we had binoculars we used to watch it."

LUKES: Did you tell anyone?

ARVID: Oh yes. Of course, we did. Because at that time it wasn't a "secret" anymore as many other people had seen something too. So that was OK. The reason for us to go further into the mountains was because there was a lot of people that were gathered on this field up here (...refers to the field "Aspåskjølen" AKA "the plateau" where many many people gathered every night. They even set up a hot-dog stand up there) at night to see if they could see anything. But every time we arrived up there, we were either too late or we went home too early so that the others saw it after we had gone home. That was the reason why we went further into the mountains to see if we could see anything there.

LUKES: How soon after did you have another sighting?

ARVID: It happened pretty quickly after the first one. Maybe no more than 2-3 days. I was up at the field "the plateau" and watched it. There I saw the "classic" sighting, it came from the south and headed north. Slowly passed us. It was usually what happened whenever you were at "the plateau", it comes from the south and headed north. But many times, when it came, it stopped in midair...then an airplane showed up...the light turned off/disappeared...when the airplane had passed the light came back on. So, it looks like it is being controlled in some sorts...

LUKES: How did that affect you?

ARVID: No... I wasn't really affected by it. Can't say I was affected by it. But since we saw it in front of the mountains and things like that, every now and then I went alone on skis and it would be nice to try and get close to it...and watch it from a close distance. So, a lot of nights I went alone on skis to see if I could see it, but I was never that lucky.

LUKES: Were you ever criticized?

ARVID: Oh yes!!! Many times! (laughing) At that time I was working in Røros, in a factory that made car windows. There was an engineer who worked there...now this had reached the media and he made fun of us almost every day for a period of time. But one time I talked to him, he asked "those lights you are seeing, are you sure it's not somebody who is staying in a snow cave, who have lights in the snow cave? " That could very well be I replied, unless you have seen a snow cave going straight up in the air (laughing). So... yes there was a lot of that. It wasn't easy in the beginning... (you can see reacting to it)

LUKES: "Describe different sightings/what did they look like"

ARVID: "Well, yes...there were different sightings but not that many. Some of them had the shape of an egg with almost a flat bottom. But there were some that looked like a cigar...they looked longer. And with that one I had an experience together with another person, who has written a book about it (probably referring to Arne With or Leif Havik). We noticed a cigar that was flying/moving, then it stopped and lowered the back end and went very close to the ground where there was snow. After a while it went back up and when it reached a certain height, it flattened out and "ppptthh" and went away (smiles and laughs)

LUKES: "What was your first thought, did you ever feel threatened."

ARVID: "No, no, no, I never felt that. Never. Because if they were up to no good, they would have done it a long time ago, because there was so much of it. (Erling responds, "yes.")

LUKES: Did you ever feel it was a good thing?

ARVID: No... but I have to say after everything I have seen, I have to believe that there is some kind of intelligence in it. I'm sure of it because there were so many things that seemed to have intelligence behind it, maybe they had some kind of communication with each other. Another experience people have had is that it turned a light on towards the ground...really big spotlights that lit up the ground...I have seen it. And that was...it was absolutely amazing it was (shaking his head).

LUKES: Spotlight? What did you feel they were looking for?

ARVID: I have absolutely no idea. If I knew, we'd come a far way. I have to say that in the beginning, after we saw it (probably referring to a sighting in the mountains with four lights) I thought it had some kind of military connection. Absolutely something to do with the military. Because at the same time that this was starting to happen, when it happened a lot, they opened up the "Hove Senter" in Hessdalen. (some kind of military base/camp) That opened up the same month. That's why I thought it could have something to do with spy equipment or such.

LUKES: Tell me about other sightings."

ARVID: Well...I have seen them many times, I sure have. But there was one special sighting when I got pretty close to it. It was me and son. We got really close. But that was down in Ålen. We could see it coming right above our heads in the forest where we were, and we could see the structure details and everything on it. I have heard of others that have seen things very similar to this. But it was a

UFOS DEJA VU

solid structure with four feet and the light we saw came from underneath it.

And the light had a structure that looked similar to a car radiator. Full of small pipes.

At this point Professor Erling Strand joins the conversation —

PROFESSOR STRAND: That was like an object?

ARVID: Yes, that was an object at around 8-9 in the evening, so it was dark and in the winter. We were waiting on snowmobiles up in the hillside near Reitan Stasjon (train station near Ålen).

LUKES: Did you ever see anything coming out of the object?

ARVID: No, no, no, it just moved very slowly in the air. Maybe 20 kilometers per hour or something like that. Not much faster. Maybe about 30-40 meters above our heads.

LUKES: Did you talk to your sister?" (his sister is Ruth Mary)

ARVID: "Yes, she has had a lot of sightings. But no, I'm not sure if I talked to her about this one. But the last observations I have had was four years ago; I must admit I haven't seen anything since then. I was on my way from Røros, late at night. I had been at a Christmas party, followed by a concert in the church. So, it must have been around 11 at night we drove past Ålen and towards Hessdalen. At the bottom of Hessdalslia (meaning the steep road up to Hessdalen). You know the phone tower (pointing, which is located a bit further north to "the plateau") was lit up. But we didn't see the light source until one of the last turns at the top of Hessdalslia, there we saw a light gliding over the ridge and toward north. It was extremely close to the ground. That is why I didn't see the light source at first because it was so low. There was a little bit of snow but not that bad so we drove up there (this is the road up to "the plateau") to see if we could see anything. But we couldn't see any traces at all in snow or nothing. But it kept moving towards north to Hessdalen.

LUKES: Did you ever find any tracks in other sightings?

ARVID: No, not really, but there was the sighting up there. (pointing towards Rogne mountain) It is said that something landed up there. That lady saw something up there. (Her name is Eli Bendos, and she lived at the bottom of Hessdalslia. So, she could look up at the north side of Rogne mountain. I believe I have something from the newspapers about this. But she hardly spoke about it afterward, she didn't want to speak out in public about it.

We used walkie-talkies, and she guided us in the mountainside, there we found tracks. It looked like something had slipped into the soil.

PROFESSOR STRAND: And this was right after she had seen this?

ARVID: (A bit unclear but I think he is saying that it hadn't been a week since she saw it.)

UFOS DEJA VU

PROFESSOR STRAND: Didn't she see something coming out it?

ARVID: Yes, something came out of it, and this "creature/alien" was moving similar to them who were on the Moon.

PROFESSOR STRAND: And then you went up there to see if there were some traces/tracks left?

ARVID: Yes, we did, in fact, see traces up there.

LUKES: Did you take any photos of it?

ARVID: No, I didn't , I didn't even have a camera myself back then. (smiling) But speaking of cameras, even though a lot of people had cameras they got so fascinated of what they saw that they forgot to use the camera until it was too late.

(I know they were taking pictures of the tracks, because I have seen them. There were also supposedly pictures of a footprint with three toes)

LUKES: Did you have problems with electricity / personal after sighting?

ARVID: No, no, never had.

PROFESSOR STRAND: You had an observation where the light came right outside your window.

ARVID: Yes, yes...I was in my bed, sleeping. Then I woke up because I felt it became brighter, and then I saw this light. I thought it was the car lights from the neighbor because their road is straight towards my house. But then I noticed the light came from above and lighting up inside and not from the road below. And then the object came gliding, not far away at all, heading towards northern the valley. It was very bright and had these spotlights on it; was a very bright light. That woke me up. I watched it as long as possible, and it was somewhere here (pointing) that last I saw of it.

PROFESSOR STRAND: Did you manage to fall asleep after that?

ARVID: Yes, that was not a problem. (laughing)

PROFESSOR STRAND: How long did it last?

ARVID: It went really slowly, about twenty to thirty kilometers per hour."

PROFESSOR STRAND: How has this affected your life in some ways since you started to see this?

ARVID: No, I can't say it has. It has been OK. But it was really exciting in the beginning because we had an apartment available, so everyone borrowed that, NRK (Norwegian Broadcasting Corporation) and many other people. And I spent a lot of time with one of the photographers, I cannot remember the name. (here he probably refers to Jon Gisle Børseth from NRK who took film/photos of the phenomenon). He was interested in it. He wanted to go into the mountains and look for it. And he got some nice pictures of it from Finsåhøgda (the beautiful mountain northwest in the valley).

PROFESSOR STRAND: NRK was up here and filmed?

UFOS DEJA VU

ARVID: Yes, many times. But the next summer his whole family, wife and kids died in a car accident. (Unclear but he says something about him not being in the valley anymore, and he hasn't had any contact with him since.)

LUKES: Has this been a positive thing for you overall?

ARVID: Well, not only positive but it has been both things, more or less. There was a period of time where there was a lot of stress with journalists, movie makers and all that you know. You never had free time...to put it like that. And when you never said "no" in the beginning, then you had it going for a while.

But have you heard about the sheriff that saw this boat burning on Lake Hessjøen? (the lake far south in the valley, very beautiful place)

PROFESSOR STRAND: No?

ARVID: Hah, no?

PROFESSOR STRAND: No, you have to tell it.

ARVID: Hahaha, well the old sheriff was really interested in this you see. He was the one who got the military up here to see if they could find it and so on. And he spent a lot of time here observing ("the plateau"). But then he told about a cabin at the north side of Hessjøen. When you come down the road towards Hessjøen. They were at this cabin, then they suddenly noticed that a boat was on fire, on an islet. And on the west side of that islet, there was a boat on fire. And he thought he should try to do something and get there and help out. But it sank and kept sinking more and more until it was gone, down in the lake.

But no one was missing, and no boat was missing or anything. And you know, many years went by, and two years ago we got a new hunting-buddy on the moose-hunting team. And I and this person stayed at the same cabin...and his family has a cabin on the opposite side, to the south. And in the evening, we started chat about different things, and we started talking about UFOs and such. "I have been wondering" he said, "when I was young and stayed at Hessjøen in the eighties. What was it that I saw out there on the lake, it was like a boat was on fire." And he told the exact same story as the sheriff told, only that he saw it from the opposite side of the lake. But they had a boat and they rowed towards it. But they didn't get to it in time, but the water was "boiling" when they got there. It was "bubbles" and they really wondered what that could have been. He is actually in the valley now.

PROFESSOR STRAND: He has a cabin at Hessjøen?"

ARVID: Yes, it's his grandfather that put up the cabin. That was amazing! And the funny thing is that one person saw it from the north side and the other from the south side. So, they had it in between them and they saw the same thing.

PROFESSOR STRAND: Didn't you have a daytime observation east of Ålen at one time? Or do I mix it with another observation?

266

UFOS DEJA VU

ARVID: Cannot say I remember that... I believe it was sometime from the mid to late eighties.

PROFESSOR STRAND: How old were you when you had your last sighting?

ARVID: Well, it was four years ago so I would have been 77 years old.

LUKES: Do you wanna see it again?

ARVID: Yes, I would like to see again. I wouldn't mind seeing it again. I would like to see it and I would like to get even closer to see it. I hope. I'm still out in the evening and nighttime when the weather is good, to see if I can see something. But you can never make an appointment with it! That's how it is.

LUKES: What has PH meant to you?

ARVID: It means a lot, that's for sure. A lot. I assume the whole thing would have been forgotten without it.

PROFESSOR STRAND: What would you tell people that has never seen it?

ARVID: Ha ha, I'll tell you. This cabin has been used a lot as a part of a rehab-center in Røros. There are a lot of patients that have been out on a trip and visiting Hessdalen. And they wanted to hear some stories about it. And of course I was the one who ended up with doing it. But what I was planning to tell them, I waited it out until they believed me. Otherwise they would think I was lying, because that's what usually happens. 14 days ago this person was renting my cabin, he was here to hunt. He was a part of a scientist team in Trondheim and we started to talk about the phenomenon. He asked if I had seen anything and I started to talk about different observations. And after I had told him different things, he said, "Now I have to sit down and think about this again before I make up my mind because of what you have told me." Because he had read and heard about it, but never believed in any of it. And I think that it's the same for many others as well. But we just have to accept that it's more between heaven and Earth than we know.

LUKES: Any other sighting you want to share?

ARVID: Well, I guess I have covered most of what I have seen. But you know, I have so many observations and I don't always remember all of them at once. But that's how it is.

* * * * *

In the end they talk about how Jon wishes he could speak English better. That he has worked with a lot of people from Scotland and so on.

After spending two weeks with the Project Hessdalen team and interviewing the locals, there is no doubt that they have experienced many perplexing things over the past four decades. It was empowering to see a team of scientists who offered support to the community and have dedicated time and resources to scientifically study anomalous light phenomena.

UFOS DEJA VU

Jon Arvid is definitely a "UFO repeater."

Above: Erica Lukes at the blue box where equipment is monitoring the activity of the lights.

Hanging out on the corner Hessdalen style.

SECTION EIGHT
ON TOP OF OLD SMOKEY

FIRST KNOWN PHOTO OF THE BROWN MOUNTAIN LIGHTS
E.M. BALL, 1929

UFOS DEJA VU

SECTION EIGHT

CHAPTER TWENTY-SEVEN
GIANT CRYSTALS AND THE MUMMIFIED MAN
By Timothy Green Beckley

CHAPTER TWENTY-EIGHT
THE BROWN MOUNTAIN LIGHTS VIEWING GUIDE
By Joshua P. Warren

UFOS DEJA VU

GIANT CRYSTALS AND A MUMMIFIED LITTLE MAN
By Timothy Green Beckley

I know of at least one UFO vantage point just about anyone with a car can travel to, and there is a good chance you will see something pretty damn "strange," as thousands of others have over a period of many, many years. This gateway to a parallel dimension is in North Carolina. It's in the middle of the Brown Mountain region, and you can stay at the nearby Holiday Inn in Morganton—about 15 miles north of an actual highway marker which has been posted by the state. It provides any visitor with the best view to look and see these mysterious lights that bob and weave across the sky, finally settling down below the ground, disappearing into the crevices.

I went there circa late Sixties with the posthumous Jim Moseley, the "trickster of UFOlogy," and the always very jolly Allen Greenfield over Christmas week. I say that because we were stuck in a rather low end cabin as I recall "singing" Christmas carols, which is remarkable because Jim was tone deaf, Allen is Jewish, and to my way of thinking its not rock and roll enough unless Elvis is in the room (maybe his spirit was).

Turning back the hands of time, I had read an article in "Argosy," which was a monthly magazine for "rugged men" with a sense of adventure (throw in one cheesecake spread of a gal in a bikini which made it as close to porn as you were likely to find in those days, outside of "Playboy)." The article had actually been written by a drinking buddy of mine, Herbert Bailey, who was listed as "scientific consultant," on the masthead of "Argosy," though he had no degree in anything but bar tending. He always carried a flask in his jacket's pocket to take the edge off. He was a good writer. Writers had a bar knuckles reputation in those days and were known for pounding them down. Though, frankly, I could never do shots and hold myself up at the typewriter. Bailey and others (like Norman Mailer) would hang out at the Lyons Den off Sheridan Square in Greenwich Village, right next to the Stonewall. The Den never got raided by the police, though the Stonewall did, because most of it's patrons were gay men, just to give you a birds eye view of an evening in the Big Apple back in "the day."

UFOS DEJA VU

There are some, I should point out, who think the Brown Mountain lights are a "natural phenomenon," but don't tell any of the locals that, because they know better. A few of them have gotten right up to the glowing globes, and they'll swear to you that these things aren't "swamp gas" (besides there is no swamp anywhere nearby) or associated with "spook lights" that so often turn out to be connected with mineral deposits in the Earth. Though, the mountains are full of a heavy deposit of quartz crystals. I've seen some of the chunks that were taken out of there. Humungous is a word that seems fitting.

Such things as car lights, or approaching landing lights from planes circling overhead, can also be ruled out, as these lights have been seen around Brown Mountain for several hundred years, perhaps going back as far as 1200. Long before the white man had reached the shores of America, the local Indians reported watching these lights, which roam about the mountain peaks without any "real" explanation.

THE LEGENDS

Various legends have sprung up about the origin of the lights. One has it that the lights are caused by the spirits of the Cherokee and Catawba braves searching the valley for their maiden lovers. It seems that the two tribes had a big battle hundreds of years ago which killed just about all the men of the two tribes. Apparently this legend does have some basis in fact, as most legends do, because within the last several hundred years at least a half-dozen Indian graves have been found in the area nearby.

However, others who have lived near Brown Mountain for as many as 75 years seem to think that there is something even more odd and peculiar than spirits at work in the valley below. According to Paul Rose, who accompanied us to a secret lookout point despite the 10-degree weather and the falling snow, they may be something from outer space.

Out of all the people living in the area, Rose, and another fellow whom we shall talk about in a minute, seem to have gotten closer to the lights than anyone else. His first sighting came when he was just a youth in about 1916. At the time it was thought that the lights might have been caused by the headlights on locomotives running through a nearby valley. However, during one rough spring all bridges were knocked out and roads were too muddy to enable cars to pass. Yet the Brown Mountain lights were seen, in greater numbers and brighter than ever before, weaving up and down over the trees on Brown Mountain.

Rose bases his opinion that they are intelligently controlled on the fact that he has seen them fighting, butting into each other and bouncing like big basketballs. He has also tracked them at speeds of almost one hundred miles per hour.

He claims that on one particular night in the late 1950s, when excitement was at an all-time high, two of these lights appeared out of the valley, approached a tower he had built for the purpose of watching them over the trees and climbed to within feet of his position.

The next day he and a friend who had been with him, both became violently ill. This led Rose to the conclusion that these lights are highly radioactive.

Another old time resident of the Brown Mountain area is Ralph Lael, who was born in Alexander county, on a small hillside farm, in 1909. He ran for Congress in 1948 and lost by a few thousand votes. At the time Jim, Allen and I were there, Ralph – now deceased – operated the "Outer Space Rock Shop Museum" on highway 181, just outside Morganton, sort of hanging over the edge of a cliff as you turned a bend in the road. It is pretty much up against a good viewing point where people go to see the lights. There was a small gravel parking lot off the winding road that led to his rather run down establishment which was more homespun than fancy. He only had a few items to sell in the store—soda pop in glass bottles, some chips and beef jerky that was homemade and not in packages. "He had a small sign that said, 'Brown Mountain Lights Museum,' and he was a nice friendly mountaineer. Maybe he wasn't as slick as the city folk who came to visit, but was certainly a rustic entrepreneur if there ever was one.

If you asked him about the Brown Mountain lights, he could chew your ear off spending all afternoon talking about them, though he would appreciate it if you also bought a copy of his small, personally autographed, 75 cent booklet on the subject. When you were finished chatting, you might pull back onto the blacktopped mountain road a bit more bewildered than when you first shook hands with him. Most of the stories he told, including those things he experienced firsthand, seemed a bit farfetched, but they certainly were entertaining. Good old Ralph would claim that he went into the mountains and had contact with these intelligent beings that were the source of the lights.

Lael claims not only to have seen the lights up close but to have communicated with them on numerous occasions as well. Deciding that the only way to uncover the source of these lights was to go into the almost impassable mountain area itself, Lael started his own investigation. Shortly after midnight he got within 100 feet of a light that had risen up from a large hole in the ground.

Within 10 or 15 minutes, the first light had been joined by as many as 20 more. Shortly after, they all took off into the timber and disappeared from Lael's view. A half-hour later others began popping up along the mountainside in a smaller valley below. One came so close, within 10 feet, that Lael felt he could have read a book by it.

Several expeditions, and months later Lael discovered that by asking the lights questions, they would answer by either moving up and down for yes or back and forth for no. After this form of communicating had been established, one of the lights led Ralph to a door, which leads inside of Brown Mountain.

UFOS DEJA VU

Once inside he was led to a room about eight feet square, the walls made of crystal "as clear as glass," enabling him to see for what seemed to be miles. Suddenly a voice said: "Do not fear; there is no danger here." The voice continued by saying that Lael has been chosen to tell the people of Earth about their true history; that man was created on another planet named Pewam which our ancestors destroyed. Pewam is now the waste of the asteroids which lie between Mars and Jupiter.

THE VOICES

The voices explained that they are not Earthbound beings and cannot eat or drink, but live on Pethine, a "gas we absorb from the light you see around us. We perish in your atmosphere or sunlight." "We live on Venus, which is a planet of pure crystal as you see surrounding you...notice that the crystal is as clear as your air. Venus is completely surrounded by water vapor about 150 miles above its surface."

In October of 1962 Lael returned to the rock, entered and was offered a ride to Venus—which he accepted. Arriving two days later on Earth's sister planet, he was introduced to men who were said to have been direct descendants of the people from the planet Pewam.

One was a rather attractive woman named Noma, who was quite beautifully dressed in a bra and panties set. While on Venus, Lael was showed what appeared to be newsreels of the destruction of Pewam as well as scenes going on back on Earth.

Lael was also warned that there are certain "forces" that could decide that man should be destroyed from the Earth. A dial was turned on the wall screen and he was shown how another planet, also the same size as our world, was rendered lifeless.

First he was shown the ice caps at this planet's North and South poles. "Then I saw a great cloud of vapor moving out from the sphere as the waters and oceans began to rush up the mountain sides and into the valleys. Trees were uprooted by the rushing waters, great licking flames of water shot up for miles from the surface of the planet. Much of it remained at great heights as it became vapor.

"As the picture drew near on the screen, I saw heaps of bodies of some type of animal like our buffalo being tossed around against 90 cliffs of the mountains and higher parts of the valley. As the axis of the sphere seemed to become perpendicular, there was so much mist that I could not see the surface. As the camera or whatever was used to make the picture drew away, the view looked like the planet Venus as we approach it in reverse."

Though Lael was very secretive about this matter, for a long while he kept a "tiny creature," or "little man" that looked "highly preserved" in a glass case in the back room of his roadside Brown Mountain Rock Shop and General Store, which

UFOS DEJA VU

he operated in order to finance his extended trips back into Brown Mountain. There he would "disappear" for weeks at a time, presumably to visit with his alien friends in their underground bunker located on the spot.

Lael hinted that this dwarflike being was actually from some distant world, had died in the area, and had been turned over to Lael because of his love and trust of those not of this place or time.

Although unbelievable as this story may seem, so are the Brown Mountain lights. Ralph Lael told us that "there are many things I have seen and heard that I cannot reveal here because of my obligations to the Brown Mountain lights. Whether you believe or disbelieve what I have told is of no importance. You and others who have read these things should have more brotherly love for the people of Earth and those of the whole universe."

While in the lair of the Visitors, Ralph was shown and handled some pretty gigantic crystals which he said they told him were a source of their power. Lael had a few of these impressive looking crystal orbs decorating his shop. Big crystals. Just like the geodes that you would find in New Ages stores, but I guess in those days we were kind of awed by their size because we didn't know too much about crystals, where we would today.

Ralph also had a petrified little man which he kept in a "coffin" beneath the counter and would pull out only for "special customers." He tried to pawn it off as a real alien, but I don't believe we fell for that one, realizing its what are called "roadside attractions," like the dozens of petrified mermaids you could find all along Route 66 that were meant to get potential customers to stop for something other than to take a piss.

But we digress a bit in order to let former Ashville, NC resident – now a La Vegas transplant – Joshua P. Warren, give us some down to earth (or more appropriately "out of this world") facts, just the facts, about the Brown Mountain Lights as he is generally recognized as a world class expert on the topic. He has not only seen the lights but photographed them as shown on the National Geographic Channel, where you don't get away with shit.

275

UFOS DEJA VU

This "lost footage" film claims tourists are being abducted and many are not being returned. But, hell, it's only science fiction, right? The long defunct Air Force's Project Blue Book launched an extensive study of The Brown Mountain Lights. The Airforce classified the Brown Mountain Lights as unexplained. Although most Project Blue Book studies were released under the Freedom Of Information Act, the documents of the Brown Mountain study remain classified and, in 1969, Project Blue Book was shut down. Watch it on Amazon Prime if you dare!

Joshua P. Warren is the main investigator of the Brown Mountain Lights, a series of ghostly lights reported sporadically for many years near Brown Mountain in North Carolina.
The lights have been seen at several locations about 60-70 miles northeast of Asheville in Burke County. Brown Mountain is located in the Pisgah National Forest. Here Joshua is spending another long and cold night searching for the unexplained. Free podcast — www.buzzsprout.com/127013/790918

UFOS DEJA VU

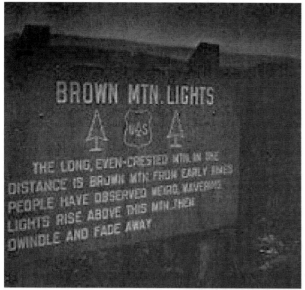

The state of North Carolina has provided a highway turn off so you can enjoy both the beauty of the mountains as well as the notorious lights. There is a sign post up ahead — you are about the enter the "Twilight Zone" — Brown Mountain style.

Spend a weekend in Ashville, NC at the Brown Mountain Festival.

Many curious tourists have spent cold and lonely nights at this overlook waiting for the Brown Mountain Lights to appear. They are rarely disappointed!

277

UFOS DEJA VU

Asheville T

Saturday, September 07, 2013

Couple Attacked By Brown I

A North Carolina couple who went to see the Brown Mountain lights on Friday night got a lot more than they bargained for. The couple said they pulled over at a look off their friends told them about and sat there for about two hours before they saw lights over a ridge several hundred yards to their west.

Then is where it gets strange. The couple told the local Sheriffs Department that a UFO craft came over the ridge and bumped the top of the couple's car several times. At one point the couple said the UFO was holding the couple's car to the

ground even though the man was attempting to back the car up and get away.

The couple said the UFO at one point hit the back top of their car so hard that it broke the back window of the car out and crushed the car down about 8 inches at the back.

The local Sheriffs Department said this was the first attack of this kind related to the Brown Mountain Lights. A spokesperson Mildred Rivers said the car received extensive damage to its top and rear areas. She says

Ren
foll
imp

The
that
rela
the
beh
of a
exp:
in li
its
beh
con
or v

It m
tote
thir
dec
mos
retu
Son

Don't be frightened; the lights can't harm you, although this newspaper clipping seems to disprove that.

UFOS DEJA VU

THE BROWN MOUNTAIN LIGHTS VIEWING GUIDE
By Joshua P. Warren

The Brown Mountain Lights are a world-famous topic of mystery and debate. For more than a century, locals, tourists, scientists and researchers have been baffled by this weird and complex phenomenon. Most people think of the lights as a wondrous, colorful display on dark ridges at night. And yet an entire subculture associates the enigma with UFOs, underground bases, conspiracies, reality warps, and high strangeness similar to the Bermuda Triangle. In this guidebook, we'll focus on the display itself.

Despite fascinating photographs of weird illuminations presented for years, some who have never seen the lights are convinced they don't even exist. There is only one way for you to know. You must venture out and look for yourself. I have been investigating the phenomenon for decades. In this guidebook, I'll give you a solid understanding of the mythos behind the lights, and tips on how to properly observe the mountain yourself. All the while, I will maintain a somewhat cautious and neutral view on what they may be. After all, they are a great mystery, and though there are many theories, no one knows for sure; hence the beauty of this rare and mystical slice of Americana.

DESCRIPTION

Usually, the Brown Mountain Lights are described as multicolored balls of light that either flare-up from one location, or move, as a group, through the trees. They are often reported flying, or floating, into the air above the mountain, as well. Even though Brown Mountain has become the focus of observation, they could more appropriately be called the "Linville Lights," since the phenomenon has been reported throughout the area, especially in and around the Linville Gorge, most often witnessed from Wiseman's View.

The lights frequently begin as a red glow, flaring into white. They can also appear as orange, blue, green or yellow. Usually, a single ball of light only lasts 6-10 seconds before vanishing. But, on rare occasion, they can last more than a minute, especially when floating into the air over the ridge. The movement of the lights is somewhat unpredictable. One orb can divide into several, the smaller ones eventually combining to form a large one again. They might seem to "ooze" around the trees and drift over the ridge; dwindle and fade away, or simply wink and vanish.

The illumination is most often witnessed from vantage points miles away. Because of such great distances, and the fleeting nature of the phenomenon, most people cannot see specific details. Over the decades, a handful of people claim to have seen the lights up-close. In 1982, one man, named Tommy Hunter, even claimed he touched one of the lights that came bobbing up to the 181 overlook. He, and other witnesses, said it was a few times larger than a basketball; a bright, yellowish color, and hovering 3 or 4 feet above the ground. When he touched it, Hunter said he received an electrical shock. The light dimmed, but did not dissipate, floating back off into the woods. Others who claim to have seen a light up-close usually give a similar description of how it appears.

Whether seen up close, or from afar, those who believe they have observed the lights cherish the spectacular and ghostly display.

HISTORY

Locals will tell you that Native American legends regarding the lights go back hundreds of years. And, in 1771, Gerard de Braham, a German engineer, recorded strange experiences while exploring the area. He was intrigued by unexplainable sounds in the mountains. In his diaries, he wonders, oddly, if the noises could be created by the spontaneous ignition of "nitrous vapors" carried by the wind. It is perhaps impossible to verify whether these early accounts are significant. However, reports of the lights exploded in the twentieth century.

On September 24, 1913, the Charlotte Observer published an article titled "No Explanation: Burke County's Mysterious Light Still Baffles Investigators." The article states, "the light rises in a southeasterly direction from the point of observation just over the lower slope of Brown Mountain, first about 7:30 P.M., again about 20 or 30 minutes later and again at 10 o'clock. It looks much like a toy fire balloon, a distinct ball, with no 'atmosphere' about it, and as nearly as the average observer can measure it, about the size of the toy balloon."

It further records, "Many have scoffed at this 'spooky' thing, and those members of the Morganton Fishing Club who first saw it more than two years ago were laughed at and accused of 'seeing things at night' as a result of a common human frailty. But as more and more persons have seen it, various attempts have been made to explain the mystery." We can therefore say, with certainty, the lights were definitely a hot topic in 1911.

After this publicity, the lights were investigated at least three times by the U.S. government: once by the U.S. Weather Service, and twice by the U.S. Geological Survey. Even the Smithsonian conducted an expedition. In a 1922 geological survey, George Mansfield studied the mountain and its weather conditions for weeks.

UFOS DEJA VU

In his official report, titled Circular 646, he stated the lights were: 47% auto headlights, 33% locomotive lights, 10% stationary lights, and 10% brush fires. Many locals felt the report was pure hogwash. The lights had supposedly been seen long before autos and locomotives. Plus, in 1916, a great flood wiped out transportation routes. There were no trains or autos in the area for more than a week. However, the lights continued to be seen.

In 1965, Ralph Lael, a local UFO researcher, published a booklet called The Brown Mountain Lights. In it, he claimed he had voluntarily cooperated in his abduction by aliens on the mountain several times. Lael's work anchored a widespread belief that the phenomenon was perhaps more extraterrestrial than terrestrial. This is reiterated in pulp magazines of the day, such as the December, 1968 issue of Argosy men's magazine. The cover story, about the lights, was titled "The UFOs You Can See Right Now."

Numerous private groups have researched the lights throughout the years. One of the most prominent investigations was done in the mid-1970s to mid-1980s by a team of scientists from the Oak Ridge National Laboratory in Oak Ridge, Tennessee called ORION (Oak Ridge Isochronous Observation Network). They, along with a group called The Enigma Project, spent years analyzing the area. Though they gathered a great deal of valuable information, they, too, were unable to conclusively solve the mystery.

Many artists have been fascinated by the mountain. The lights have been featured in numerous books on the unexplained. North Carolina authors John Harden, Nancy Roberts, and John Parris have all included the phenomenon in their works. In the early 1960s, song writer Scott Wiseman's "The Legend of the Brown Mountain Lights" became a bluegrass hit, performed by Tommy Faile. It has been rerecorded by numerous artists.

The Brown Mountain Lights have also been popular on television. They were featured in a May 9, 1999 episode of The X-Files called "Field Trip." They have been showcased on the Travel Channel, Discovery Channel, and National Geographic Channel, and continue to pop up on documentaries around the world.

During a series of trips in November of 2000, the first good video footage of the Brown Mountain Lights was captured, using Sony night-shot, by Brian Irish, imaging specialist for my research team, L.E.M.U.R. (League of Energy Materialization & Unexplained phenomena Research). Brian and I watched a particularly amazing display on a cold, windy, rainy night. When many people think of the Brown Mountain Lights, the image that comes to mind is the footage captured those dreary fall evenings.

In 2012, the Burke County Tourism Development Authority produced the first official Brown Mountain Lights Symposium to discuss the mystery, drawing a crowd of hundreds from around the country. The event was so popular a second one was held later that year. All viewpoints were given a stage, and diverse opinions on the lights were as passionate as ever.

LEGENDS

There are countless legends surrounding the lights. One of the most popular was featured in the "Legend of the Brown Mountain Lights" bluegrass song. The author, Scott Wiseman, was the nephew of Fate Wiseman, after whom Wiseman's View was named. Here is that version of the tale:

Brown Mountain was named after a plantation owner who lived in the area in the 1800s. He was kind to his slaves. One night, he ventured onto the mountain to hunt. When he did not return, one of his slaves took a lantern and scoured the ridge for him. He, too, was never seen again. Today, you can still see the "faithful old slave's" lantern burning as his spirit still searches for his lost master. Of course, the mountain frequently produces multiple lights at the same time. However, Wiseman must have liked the image of a single, devoted spirit. The song using this legend was performed at the Grand Ole Opry and became quite a hit. For a while after its release, light watching reached a peak.

Another popular tale, which dates back to the year 1200, involves a vicious battle on the ridge between the Cherokee and Catawba Indians. Some of the best warriors died. That night, after the fight, the Indian maidens lit torches and scoured the ridge for bodies. The mournful scene was so tragic and intense it still haunts the mountain.

Others tell of a man who murdered his wife and child and secretly buried them on Brown Mountain. Shortly thereafter, the lights began appearing over their hidden graves. Locals were drawn to investigate the illumination and discovered the bodies. However, the murderer escaped and was never seen nor heard from again.

Even though most of the legends portray the lights as ghosts, tales of UFOs, aliens, interdimensional beings, little people, fairies and such have become widespread since the 1960s. There are some who believe the lights might even be conscious beings of energy who live inside the mountain. Brown Mountain has inspired more myths, folk tales, and superstitions than one can easily document. It's certainly easy to understand why.

THE MOUNTAIN

Brown Mountain is one of the last formations as the Blue Ridge Mountains submit to the flat Piedmont region. It stands in the Pisgah National Forest, near Morganton, on the border of Burke and Caldwell Counties, in western North Carolina. Brown Mountain is a small peak with a low-lying ridge that runs along the horizon beside it. The peak is 2,750 feet, while the ridge stands around 2,600 feet. On the surface, they appear as ordinary as any rise of land in the surrounding mountains.

A portion of the mountain is home to the Pisgah National Forest's only ATV

UFOS DEJA VU

(All-Terrain Vehicle) trails. These extremely rugged paths loop for miles on and around the mountain. It's one of the country's most exhilarating parks for dirt bike and 4-wheeler enthusiasts.

Hikers can enjoy the rest of the area, with the exception of a patch of private property. There is a Forest Service road that leads toward the ridge. However, it's blocked by a locked gate used only by the Forest Service. Civilians can request a free permit and key from the Nebo, North Carolina Forest Service station to unlock the gate and drive closer to the mountain for camping.

The backside of the mountain is quite lush. Scenic streams rush through the crags, cutting mighty gapes through the mountain boulders. There are ferns, mushrooms, mosses, and other signs of moist life. However, as one ascends the mountain, the moisture dwindles. Many of the trees on the mountainside look sparse and dead, perhaps suffering from acid rain.

The ridge is primarily composed of ordinary "cranberry granite." It contains sandstone, quartz, and mica. Iron has also been found on the mountain. The area around Brown Mountain is a black bear reserve, and it's also well known for copperheads, one of the most deadly snakes in the country. The Brown Mountain area is extremely rugged, thus hikers and bikers in the area have died from accidents. The side of the ridge is covered with slick rock faces, especially treacherous when covered with water or fallen leaves. One of the biggest reasons the lights are still a mystery is because it's so dangerous and difficult to navigate the ridge and surrounding land. There are also some small caves and holes around the mountain. It's noteworthy that Brown Mountain is a short distance from the Linville Caverns, the only "show cave" in the state of North Carolina. Interestingly, the ridge is almost completely encircled by thrust faults. This fact plays into some of the theories regarding the lights being related to geologic movement.

THEORIES

There are places on earth where unexplained, ghostly lights sometimes appear at night. They're generally called spooklights, will-o-the-wisps, foxfire, jack-o'-lanterns, ignis fatuus (Latin for Fool's Fire) or simply, earth lights. Famous examples are the Marfa Lights in Texas and the Hessdalen Lights in Norway.

Some claim that "swamp gas" (principally methane) released by dying plant and animal matter spontaneously ignites and creates the effects. However, most enduring spooklights do not behave in a manner consistent with traditional gaseous activity.

The Brown Mountain Lights occur on the side of a mountain where no swampy areas exist. However, they don't appear "gaseous," anyway. When gas is released into the air, it spreads and diffuses into the atmosphere. The Brown Mountain Lights appear to be self-contained, concentrated balls of light which can maneuver the mountain. While traveling, and clearly not attached to a stationary "fuel port," they can continue to "burn" for a minute or more. The lights can also be extraordinarily bright (even when viewed many miles away); seemingly far too bright for known natural gas to produce. They frequently appear when the conditions are dry.

Why wouldn't balls of ignited gas burn up the mountain as they move through the trees? They've never been known to start a fire. It is true that, in the 1960s, a group of scientists on the mountain first reported feeling faint when the lights appeared, as if they were exposed to gas. And this still happens from time to time, though there is no evidence of such a gas that could affect people so far away. However, some people may be sensitive to intense blasts of electromagnetic radiation.

A strong theory is that the lights may be an electrically-charged plasma (a state of matter like a candle flame), similar to ball lightning. The formation of such manifestations in nature is, in itself, still largely a mystery to current scientists. In 2004, my team and I released a report on the cause of the Brown Mountain Lights.

By reproducing conditions found on the mountain, we were able to create a similar phenomenon on a miniature scale in our laboratory. We concluded most of the lights are indeed a kind of ball lightning produced by some special characteristics of the mountain. Essentially, we feel the mountain is sort of a large, natural capacitor—something that stores up electrical energy over time—then releases it at a critical moment, where intersecting discharges merge to create spinning points of light. If so, this might explain why a man like Tommy Hunter reported an uncomfortable, but relatively harmless, shock when he touched a light. In 2000, photographer Mark-Ellis Bennett shot two rolls of infrared film, through a visually-opaque filter, as the lights appeared in the distance. Both rolls were entirely overexposed, seeming to indicate a massive amount of electromagnetic energy being produced. Additionally, on many occasions over the years, our team measured disturbances in the radio and electrostatic environment when strange lights were spotted. Those who have seen a light up-close sometimes say it will tend to follow you if you walk away, or move away if you move toward it. If true, this could be due to your own body's electrical field interacting with a light's, similar to the field between two magnets.

Our findings made the cover of the October, 2004, Electric Spacecraft journal, and have been praised by numerous researchers, including a plasma scientist at the Naval Research Laboratory in Washington, DC. In 2010, our work was documented for the National Geographic Channel. Footage of a weird light was captured by our researcher Dean Warsing. The footage was sent to the Princeton Optics Lab for analysis. Their conclusion: "We basically went through all of the wavelengths starting from ultraviolet all the way to the near infrared and were not able to replicate the blue color... It is unexplained at this point."

UFOS DEJA VU

The Brevard Fault, a major force in shaping the Blue Ridge Mountains, does run through the vicinity of the ridge. However, according to geologists, it hasn't moved significantly in 185 million years. But the smaller faults around the mountain are thrust faults that are capable of sliding, perhaps as warm daytime temperatures morph into a chilly night (most drastically in the fall). Brown Mountain contains lots of quartz. Might local fault movement sometimes apply pressure to the quartz crystals, producing electricity? Great pressure applied to all rock, but especially quartz, produces electricity. This is called the piezoelectric effect, and might explain how various substances could become "excited" or ignited by energy on the mountain... but then, what substance exactly?

ORION scientists detonated dynamite on the mountain in an attempt to stimulate the lights. They had some success creating odd flashes of light, but nothing as substantial as the prominent displays reported. During that same period ORION used a spectroscope to analyze lights on the ridge, but say they found nothing distinguishable from ordinary household lights.

Others have blamed the illuminations on atmospheric reflections and refractions of both artificial and natural sources of light. Some researchers have even wondered if starlight could be refracted in some unusual way. However, evidence for these theories has never been presented. Besides, many research teams like ORION and L.E.M.U.R. have documented anomalous energy fields when the lights appear. These include readings on standard EMF meters, IR scanners, and even Geiger counters (not due to ionizing radiation, but because of high electrostatic charge affecting the tube). Optical illusions could not create these types of objective fields, nor could conventional man-made lights.

Lights from campers and off road vehicles are commonly mistaken for paranormal illuminations. However, the ORV park was only installed in the 1980s, and the lights have been seen for at least the past century. In fact, the scientists from ORION conducted most of their research in the 1970s and early 80s, before the ORV park was constructed.

Of course, the most fascinating theories involve the true fringes of the paranormal and metaphysical. Many believe the lights are the product of UFO activity. One local, a U.S. Congressional candidate named Ralph Lael, especially championed this connection in the 1960s. He claimed he had learned to communicate with the lights via a telepathic mental connection, and he was directed to a hidden cave nearby, filled with crystals. There, humanoid aliens from the planet Pewam took him into space on multiple occasions, advising on why humans must improve to save the planet. Lael even kept a small, mummified body in the back of his local shop, The Outer Space Rock Shop Museum. He claimed it was the body of an alien! Upon Lael's death, no one is sure what happened to this bizarre little creature.

Along these same lines, Brown Mountain is specifically noted as an under-

UFOS DEJA VU

ground alien base in the book titled, fittingly enough, Underground Alien Bases, by Commander X. It was first published in 1990, but has since been reprinted. The publisher of this book is New Yorker, Timothy Green

Over the years, hunters have reported military exercises buzzing around the mountain and Humvees that vanish into the forest, as if they disappeared down a tunnel. Many researchers and photographers claim to have been harassed by all kinds of shady officials and "men in black" for snooping around. Why would the military be interested in Brown Mountain? Perhaps they are interested in weaponizing the energy that creates the lights, especially if it is a powerful plasma. Or if the military is in cahoots with aliens, this might be a perfect rendezvous. Or maybe Brown Mountain is simply a portal that distorts the laws of physics, and makes contact with other, advanced realms easier.

The idea that Brown Mountain is a psychic vortex, as Page Bryant claimed in her 1994 book, The Spiritual Reawakening of the Great Smoky Mountains, opens a supernatural can of worms. Is Brown Mountain like a miniature Bermuda Triangle, swirling with cosmic energy? If so, it would explain why airplane pilots report spinning compasses as they fly over. And it might also give us insight on the time warps, strange visions and sensations visitors have. Canadian scientist Michael Persinger has demonstrated, in lab experiments, that various electromagnetic frequencies can induce paranormal experiences when directed at the human brain.

There are countless stories of beings of light, little people, glowing fairies, a creepy "pumpkin man," and interdimensional activity around Brown Mountain. Some have even attributed the lights to giant fireflies or weird worms with "flaming blowholes!" It is therefore easy to see that whether one prefers a traditionally-scientific view, or a more imaginative one, there is a vast spectrum of hypotheses, theories and opinions on what the Brown Mountain Lights may be. Maybe they are all correct. Perhaps a place that naturally produces so much energy also spawns a plethora of other paranormal events, whether considered subjective or objectively measurable.

WHEN AND HOW TO OBSERVE

Brown Mountain, seated in the Pisgah National Forest, is on the border of Burke and Caldwell counties in Western North Carolina.

Enjoy the scenic route: Take 1-40 and exit at Old Fort (Exit 72). Continue straight through the town of Old Fort. After 11.5 miles (from exit) you'll reach an intersection and traffic light with a Wal-Mart on the left. Take a left there onto 221 North, and travel 29.4 miles. There, turn right (beside a gas station) onto Highway 181 and travel 9.5 miles. At mile marker 20, you'll find a pull-off on the left side of the road. Brown Mountain is the long, low-lying ridge straight across from the overlook, 3.5 miles away. The approximate travel time is 1 hour 15 minutes from the Old Fort exit.

UFOS DEJA VU

Or, try the more direct route: From 1-40, take Exit 100 (Morganton). After coming off the exit ramp, go 3.2 miles toward Morganton (left if you're traveling East and right if you're traveling West), and take a left at the stop light onto 181 North. Travel 18.8 miles: You'll find the pull-off on the right at mile marker 20. Approximate travel time is 30 minutes from the exit.

You can also see the lights from other various points in the area, like Wiseman's View (on the edge of the Linville Gorge), or Grandfather Mountain. However, the overlook on 181 is by far the best spot to see Brown Mountain.

Regardless of the conditions, clear or cloudy, rainy or dry, new moon or full, the lights, when they appear, are usually bright enough to be seen. Though rare, the phenomenon apparently occurs now as much as ever. However, it certainly does not occur every night. Though the lights have been reported in virtually all weather conditions and at all times of year, they are more prominent in the fall. This may simply be due to better visibility when the leaves are off the trees. Or perhaps the additional acid, added to the ground water, from the decaying leaves, provides a nice conductive liquid, or electrolyte, for the ridge's electrical activity. In fact, numerous campers on, or near, Brown Mountain, have seen balls of light bouncing around streams, and even bobbing up or down waterfalls.

Do not be misled by lights from airplanes, towers, or the nearby towns of Lenoir and Morganton. The paranormal lights are extremely pronounced and cannot be confused with these conventional and consistent illuminations. They often move in strange ways and can change color before your eyes, unlike the stationary lights from normal sources, or car lights that always move on the same route.

As the product of a complicated natural mechanism, though no one can say for certain when they will appear, my and my team's research has indicated you have the highest chances of seeing the lights when the following four conditions are met:

1. It is autumn: due to the impact of the leaves; plus the temperature differential between night and day may enhance the mountain's movement, expanding and contracting to stir up energy production.

2. It is either raining or a rain has recently occurred: The water running on and through the mountain may build up electrical charges.

3. There is extra carbon (smoke) in the air, perhaps from campers or a forest fire: Such debris in the air may provide a fuel for the bright lights to consume as they "burn."

4. The Kp-index reads 5 or above: The Kp-index is a measurement of how disturbed the earth's overall magnetic field is. This field, or magnetosphere, warps drastically, moment by moment, in response to solar flares and space weather. When the field is disturbed, it induces more current flow in the planet, and enhances celestial effects like the aurora borealis and aurora australis. For a live

UFOS DEJA VU

update on the Kp-index, do an online search or simply visit:

www.BrownMountainLights.com for a readout. A 5 or above indicates a storm, and is represented by a red bar.

If you are lucky enough to see the Brown Mountain Lights, please be sure to contact us through the site to share your experiences and the environmental conditions when you saw them! Better yet, come visit us at the Asheville Mystery Museum to see our Brown Mountain Lights display and tell your story. See: www.HauntedAsheville.com

CONCLUSION

Whatever they are, the Brown Mountain Lights remind us that earth is a powerful and dynamic machine. NASA has shown us that the very gravitational field is an inconsistent and irregular force about its surface, and enhanced imaging techniques reveal spectacular, eerie flashes of light and electricity in and all around the planet. Each second, somewhere on the globe, bolts of lightning strike the ground, banging the earth like a giant drum, and strumming out cosmic harmonies. Whether the lights begin with something geologic, atmospheric, or truly interdimensional, there is a greater story here...

If we strip away all the opinions, we ultimately have a place where, for over a century, humans from all walks of life have essentially described something simple: weird lights on a dark mountainside. The geologists focus on piezoelectricity. The chemists focus on gas. The astronomers focus on optical illusions. The physicists focus on plasma. The spiritualists focus on ghosts. The UFO hunters focus on flying saucers. And the conspiracy buffs focus on mind-bending plots. Yet all the while, the songwriters, poets, and artists are just as inspired by their own, personal multicolored visions. The mountain is a phantasmagoria of strange tales, but ultimately, it is a vast blank slate. It is the tracing board for all manner of human adventure, exploration, wonder and imagination. Regardless of what they are, the mere concept of the Brown Mountain Lights makes them important. And in this internet age, when the whole universe seems right at our fingertips, how refreshing it is to have a good, old-fashioned mystery still quietly waiting to be solved, right here, in our own backyards.

ABOUT THE AUTHOR

Joshua P. Warren was born in Asheville, North Carolina, and began studying the Brown Mountain Lights as a teenager. He and his team spent over a decade camping on the site and bringing a variety of scientists to supervise their work and share opinions.

Warren is currently the author of fifteen books, including Haunted Asheville, the first book of Asheville ghost stories ever written. He owns the popular Haunted Asheville ghost tours and the Asheville Mystery Museum. Warren frequently appears on television, and is known for his work on the Travel Channel, History Chan-

UFOS DEJA VU

nel, Discovery Channel, National Geographic Channel, Animal Planet, TLC and SyFy.

He hosts the Speaking of Strange radio program Saturday nights, and is a correspondent for Coast to Coast AM, the largest overnight radio show in the country. You can learn more about Warren, and how you can participate in his expeditions around the world, at: www.JoshuaPWarren.com

It's the oldest & original ghost tour in Asheville, North Carolina, heart of the oldest mountains in North America And we take you INTO the Asheville Mystery Museum in the basement of the Asheville Masonic Temple!

Learn about:

· The Pink Lady who haunts the Grove Park Inn

· Chilling apparitions from a suicide at Helen's Bridge

· A young lady brutally murdered in the Battery Park Hotel

· Spirits from the madman who executed our city's largest killing spree

· A body entombed in the wall of St. Lawrence, and much more.

· PLUS, see Art Bell's Alien Statue that reportedly "comes to life" at night!

· Not only will you experience Asheville's dark underbelly, but you'll also learn about ghost hunting yourself. It's the original adventure produced by Asheville's original ghost expert. You'll LOVE IT or your money back! See why we've been rated #1 on Trip Advisor.

$20 per person! · RESERVATIONS REQUIRED www.HauntedAsheville.com

(828) 398-4678 80 Broadway Street · Asheville, NC 28801

Paid parking is available next to the building, or you can park on the street for free after 6pm.

Josh Warren's Brown Mountain Lights Viewing Guide

SECTION NINE
ENTERING THE TWILIGHT ZONE

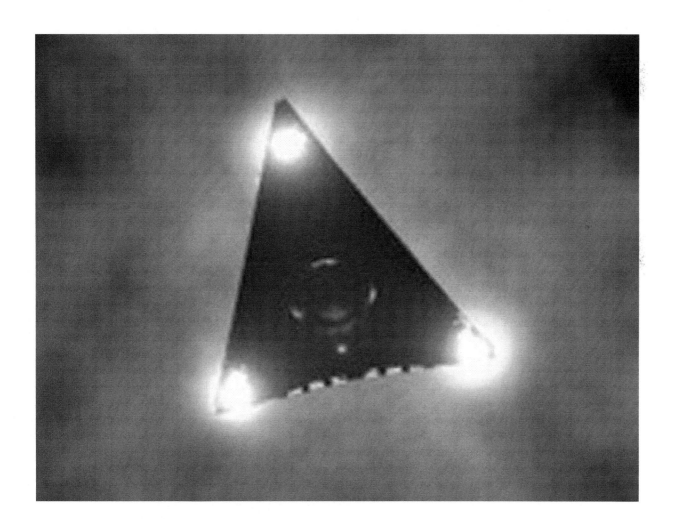

UFOS DEJA VU

SECTION NINE

CHAPTER TWENTY-NINE
THE NEW ENGLAND TRIANGLE
By Timothy Green Beckley

CHAPTER THIRTY
SOME CALL IT THE "THING"
By Timothy Green Beckley

CHAPTER THIRTY-ONE
ADJUNTAS—A PUERTO RICAN GATE TO THE UNKNOWN
By Scott Corrales

UFOS DEJA VU

THE NEW ENGLAND "TRIANGLE"
By Timothy Green Beckley

We can't say how much of a "triangle" it is or where it's points extend to. Its pretty much accepted as fact that UFOs are a "sure bet" throughout New England.

In fact, as far as can be determined the first sighting was in this neck of the woods ("woods" would have been all there was around at the time). And would you believe (sure you would) that it included a shapeshifting creature and a possible abduction? Amen!

The incident dates back to 1639, when Massachusetts Bay Colony cofounder and governor, John Winthrop recorded a secondhand observation of unidentified objects in the sky over Boston. In his diary entry of March 1 that year, Winthrop wrote that a "sober, discreet man" named James Everell was rowing a boat up the Muddy River at night when he saw a "great light" in the sky. Winthrop reports that "when it stood still, it flamed up, and was about three yards square; when it ran, it was contracted into the figure of a swine: it ran as swift as an arrow towards Charlton [Charlestown] and so up and down about two or three hours." By the time the lights moved away, Everell and his boat mates had been delivered one mile upstream, although they had no memory of how.

Just to fool around, when I'm on a talk show and want to get all serious about the strange phenomena reported throughout New England, I read to the listeners the following, without telling them from which book I am quoting:

"A drowsy, dreamy influence seems to hang over the land, and to pervade the very atmosphere. Some say that the place was bewitched by a high German doctor during the early days of the settlement; others, that an old Indian chief, the prophet or wizard of his tribe, held his powwows there before the country was discovered by Master Hendrick Hudson. Certainly, the place still continues under the sway of some witching power that holds a spell over the minds of the good people, causing them to walk in a continual reverie. They are given to all kinds of marvelous beliefs, are subject to trances and visions, and frequently see strange sights and hear music and voices in the air. The whole neighborhood abounds with local tales, haunted spots, and twilight superstitions; stars shoot and meteors

UFOS DEJA VU

glare oftener across the valley than in any other part of the country. . ."

The quote, in case you were not an English lit major, is from Washington Irving"s "The Legend Of Sleepy Hollow." You know the story about Ikabod Crane and the Headless Horseman, and if you didn't, you can tell from the above paragraph that Irving could just as well be talking about events along the New England corridor as if it were "Breaking News," and not historically mandated.

OLD CAPE COD

Diana Desimone, a Massachusetts researcher for the Mutual UFO Network, says Cape Cod is now the most UFO inundated part of the Bay State. Fishermen returning to Provincetown said they glimpsed a circle shaped, lighted object in the sky near a lighthouse in 1989. Desimone says she often hears of unexplained amber colored lights cruising over the moors or offshore. But, she allows, they could be planes headed to Logan International Airport in Boston.

Observers think otherwise! UFOs have also made repeat appearances in the Winsted-Torrington area. Police, campers, nurses, teenagers and small boys have all reported seeing strange, brightly lit objects in the sky over the last 25 years. And then there's the 1986 incident — more than 200 people watching as an airborne strip of lights illuminated the night sky over Highland Lake in Winsted.

One witness at Dennis, Cape Cod, recalls an incident from 1966 after his family visited one of his father's Army buddies on the July 3 or 4, according to testimony in Case 68411 from the Mutual UFO Network (MUFON) witness reporting database. "The man had a son my age (14) so we buddied up for the day, met up with a couple of girls he knew and all spent the day together," the witness stated. "That evening about 30 to 40 people of all ages assembled on the small beach at the lake to watch a fireworks display from a man who brought fireworks in from New Hampshire. To hold private fireworks events in Massachusetts was illegal at that time.

The UFO appeared nearby as the group watched the fireworks.

"As Roman candles were shooting into the air, I noticed that people were all looking away from the fireworks and pointing at something across the lake. I followed their pointing fingers and then saw the UFO."

New England's portal seems to remain open pretty much all the time. It is fairly easy to run into people who have observed something unusual in the sky as recently as "last week." We're talking triangles, boomerangs, odd shaped ships with pipes and metal grills hanging all over the place. Check out one of Linda Zimmerman's books on the Hudson Valley flap and tune into a conversation which we had with her on Exploring The Bizarre, archived on Mr. UFOs Secret Files on YouTube.com — just type in her name on our channel's search bar, where the little magnifying glass appears..

* * * * *

UFOS DEJA VU

The "most memorable — modern day — flap" in what I have dubbed the "Triangle" dates back to the 1980's. At that time, dozens of frightened and utterly perplexed witnesses had reported seeing a strange triangular-shaped object as large as—or larger—than a football field, gliding slowly and at times hovering in the evening sky. For the most part, the craft was said to move silently, and was made up of a row of multicolored lights that would turn on and off at random intervals silhouetted against a magnetic-gray colored hull.

The truth is that several of those who got an exceptionally close look at the ship remarked that it slightly resembled Star Trek's Enterprise. But, believe it or not, this wave of sightings can in no way be compared to something out of a science fiction movie or TV series, for there are now many—numbering well into thousands—whose lives have been totally altered after having encountered a visitor from the deep unknown.

"This is the biggest mystery I've ever been confronted with," commented astronomer and UFO investigator Dr. J. Allen Hynek as he stood on stage smoking his pipe in front of an auditorium filled to overflowing with those who had come to the high school in the tiny upstate community of Brewster, New York, desperately searching for answers...answers to an enigma which the authorities were proclaiming to be a hoax of gigantic proportions. More than a thousand area residents waited with bated breath for an opportunity to learn from one of the country's leading UFOIogists what it was they had gotten more than a glimpse of in the sky.

Lawyer Peter Gersten and teacher Philip Imbrogno organized the conference after it became clear that the public would no longer take the government's constant denials for what they had seen. Imbrogno, who is a field investigator for Hynek's Center for UFO Studies says that he has documented evidence that two sightings in the area—those of June 11 and July 12, 1964—were witnessed by upwards of 6,000 individuals, and that the total number of observers over a period of a year and a half may come close to the ten thousand mark.

On both of the above dates, motorists piled out of their cars along major roadways to watch as the sleekly constructed ship glided overhead without making a sound. With heads turned toward the night sky, they watched in awe as the spectre descended to treetop level. One police officer told how the UFO came to within 60 feet of his patrol car, appearing to play a sort of cat and mouse game with his vehicle, before reversing direction, and passing directly over the roof of his auto.

Though authorities were quick to arrive at a satisfactory (to them) explanation, stating that the sightings were caused by a formation of ultra-lights—one-man open aircraft-flying after curfew. But, try as they might, nobody who had been privileged to witness the phenomenon was buying this answer completely as they instinctively knew something much more "unearthly" was probably involved.

"The high school in Brewster was picked for the public forum because a good many of the sightings had been made by residents of this suburban community," Philip Imbrogno explained as the conference started. "Furthermore, the caliber of the witnesses here is much higher than usual," stressed the professional educator, "and in addition, unlike in the vast majority of cases, this UFO has stuck around for over a year and a half, not having vanished into thin air right away as in most instances. This factor has given us added hope that something concrete can be arrived at in the way of an explanation."

The highlight of the conference itself, was the testimony of a dozen witnesses who came forward to publicly relate their experiences; a good deal of what they said providing evidence that the craft in question is being intelligently controlled.

IT COULD READ MY MIND

Putnam county mental health services employee Monique O'Driscoll feels strongly that whoever was inside the craft had the ability to read her mind. Speaking into the onstage microphone, the articulate witness told about what transpired on the evening of February 26,1964, after she had finished having dinner at her mother's house.

"My mother lives in the town of Kent near Route 52. My daughter was with me at the time. We left my mother's house at 8:30. It was dark. We went down the road to the first stop sign that reads Farmers Mills Road. "At that point, my daughter said to me, 'Mom, look at those lights up on that hill.' I said, 'Oh boy, someone must be having a real party up there.' It was like strobe lights flashing more than you would see at a discoteque. As I looked at it, it was moving. My daughter said, 'That's not a house up there, it's only a hill.' Then I realized there were no houses up there.

"I made a right hand turn to go towards home. I don't know if I followed it or it followed me. The object curved around the road, which was a winding road. It was quite low. When the object came about a mile further down the road, it had to lift itself up to go above the church steeple. At this point my daughter said, 'It's going to go over to White Pond.' It did go into White Pond and I followed it. There were no leaves on the trees, so I could easily see it through the trees.

"About a quarter of a mile down, on this dirt road called White Pond, it went in front of my car. I could see its underside. The lights were so brilliant, that I couldn't even describe them to you unless you have seen them. There were reds, blues and gold. I got out of the car and walked behind a birch tree which was in the way. That's how close this thing was.

"Now, I had a closer view of the object, it was close enough to hit with a baseball. The V part was away from me. The lights were going from left to right in an almost computer fashion. I stared at it hypnotized, wondering how It could be so big and yet not make a sound. I could see the solid structure around it.

"After a few minutes, it started floating away. I said to myself, 'Oh, please don't go away. I just want to look at you! It might sound crazy but l think that it heard me, because at the moment that I said that, it stopped and did a complete turn around. Now the V faced me and began to move towards me.

"While this was happening, my daughter stayed in the car. She was terrified and kept yelling, 'Mommy, come back in the car. They're going to take you away.' We had a CB radio in the car and she tried to make contact. All we got was static. We couldn't contact anyone, even though I could hear the truckers on Route 84."

HYPNOTIZED

"Meanwhile, I could not take my eyes off the object. I can't explain why I wasn't afraid when it started to move towards me, it had a good feeling about it. I was concerned about my daughter, when she started screaming her lungs out.

"As it started towards me, the one big headlight in front was hitting me in the eyes. It was an amber colored light, slowly floating towards me. When it got really close, I started feeling a little uncomfortable and I took a couple of steps back. When I did this it stopped, as if sensing that I was afraid. It hovered in midair for a while. I said, 'Damn it, I wish I hadn't backed up.'

"As it began to float away, I returned to my car. My daughter said, 'You're crazy, Mom. They're going to take you. Then, what do I have?' I said, Well, I have a life insurance policy.' I backed out of the road and started following it. I followed it up Farmers Mills Road. Then my daughter said, 'It's going to go to Gypsy Trails Road,' This was the second time that my daughter predicted which way it was going to go. Then it kept going, right beyond Gypsy Trails to where there are two houses. It stopped by the second house, which has a plateau.

"At this point, I pulled my car over again and got out. I blew the horn realty loud. The people in the house were looking out their front window at me. They must have thought I was really crazy, trying to get their attention. They did not come out. This thing was right above their house, about 15 feet above the telephone lines. It was so big that, even though it was over their house, it was over me. I was looking up at it.

"I was looking at the underbelly. It had a solid dark grey metallic grill type of frame. It had two grill things hanging down from it. There were no lights underneath it. The lights on top were still going around.

"A car want by with a couple of fellows in it. They had seen it and were giggling. They kept going. Then I waved down a car with a couple in it l said, you have to took at this thing.' I knew the person and she said, 'Don't even bother telling anybody. No one is going to believe you.' After a while, it started moving away and I got back in my car and followed it. I was really sad. I didn't want it to go. I went around the curve of Farmers Mills Road and about a mile and a half up the road it suddenly was gone.

"I went home and told my mother about it. I immediately got my sketch pad

UFOS DEJA VU

out and started drawing it from different angles. Even if I had not sketched it, I would never forget what I saw."

MISSING TIME

"About a month later, I was lying in bed, watching the Odd Couple on TV, unable to fall asleep. My bed is right next to a picture window. After the show, I turned off the TV and closed my eyes. Something told me to look out the window.

"There were five lights right outside my window, 300 yards in the air. I got my camera and went outside. I couldn't get a good picture because the telephone wires were in the way. As soon as I put the camera to my eyes, the lights went out.

"When I went back inside, I was unable to sleep, so I took the dog for a walk down by the lake. It was only a block away, but on the way back I lost so much strength that I hardly made it back home. I left for the lake before midnight, but I didn't get back until after one. I don't know what happened.

"On March 25th of 1984 I had my next sighting. It was my mother's birthday and I went to her house after work. My brother was there with his girlfriend, my daughter was there, and my mother and myself. The same ship that I saw February of last year was there; we all ran out and saw it. It was boomerang-shaped with white lights. As it hovered, all of a sudden, the lights turned red. Then it disappeared.

"The following Sunday I went there with my camera. I do photography on the side. I knew that it was going to come back, so I went back there. I took four pictures with my camera. None of them came out. I know I had enough light. The other pictures on the same film developed. "

SPIRITUAL ENCOUNTER

Another Putnam County employee, Dennis Sant, was also adamant about what he saw, feeling convinced that the authorities were attempting to pull the wool over everyone's eyes with explanations which just did not fit the facts.

"On March 17th, I was taking my children home from a youth group at the Baptist Church. On the way home I noticed an object with a string of lights sort of like a tractor trailer, parked in the middle of the sky, hanging over my backyard. I stopped the van and the children got out to watch this thing hovering over my yard. When I realized it was over my backyard, I drove the van back to my driveway and we all got out.

"When we got there, the UFO was gone and the sky was pitch black. So I took the children inside and prepared them for bed. While I was discussing with them what we had seen, I had a strong urge to go back outside. I had a gut feeling that there was something out there.

"I walked out the door and saw a large object hovering over route 84. It had a string of lights and hovered about 40 feet above a tractor trailer truck. I watched it for a few minutes, then I ran inside to get the rest of my family, my dad and my

298

children.

"We ail went outside and watched the object from the same spot for about ten minutes. As we talked about what was going on, I felt envious of the truckers, who were now pulling off to the side of the road, getting a closer view of the object. I had an extreme desire to be closer to the object.

"At this point, the object turned around 180 degrees. It was now a V-shaped object and it began floating towards my home. It came to a halt about 100 yards away and about 40 or 50 feet above the trees. It hovered over the field for four or five minutes, until my dad and I walked across the road towards the object.

"From what I had read about UFOs, I had always believed that I would be frightened if I ever encountered one. I want you to know it had nothing to do with being frightened or my running away. To me, it was an emotional, even a spiritual encounter.

"As we crossed the street, we saw the lights—red, amber, green, and a large white light in the middle. The lights became more intense, illuminating the area, which was a swamp and the road. The children ran back into the house. Then the object backed around my property and my neighbor's yard. Then it came back to hover in the original spot, where we had first seen it over my backyard.

"At this point, I had an emotional rather than a visual sighting. This was something I had never seen before in my entire life. We, who were brought up here in Putnam County and Brewster, are used to seeing planes in holding patterns. We are used to seeing planes from Kennedy to La Guardia airport. We are used to the small planes from all around the area We were used to seeing things in the sky but we weren't used to seeing this.'

"In my mind I imagined the object having conflict with land and I became very excited, as well as, nervous and anxious. At this point, the object began to move away from my yard in a northwestern direction. At this point my neighbors began to come out to witness what they believed was a large plane crashing in the trees.

"I did not come forward right away with my sighting. One reason was that it was a very personal sighting for me and my family. Another reason was that I didn't know what reaction the other people would have. This was a very early sighting, before the mass sightings began. At this time I thought it was the only one to have seen this thing.

"After contacting the sheriffs department, I found out that on that very night, just north of me, many cars were pulled over to the side of the road, as the object was seen over the entire area. So I gave an interview to a newspaper reporter and said, 'Please, don't use my name, but I want everyone to know that there is something in the sky above us.'

'I don't know if this UFO was an object from another world, but I think, as a people, we ought to ask, 'What in the world is hanging over our heads and our community?' We need to ask some hard questions of our local and federal officials about these objects, which are invading our airspace. If you want to tell me it's ultralights, my goodness, we are talking about five or six hundred sightings. I think we need to put these people away."

FROM VIETNAM TO BREWSTER

After spending time in service to his country, William Sockui has every reason to be upset about the way Uncle Sam has attempted to treat the entire situation. Before discussing his most recent experience, the witness described his first UFO sighting which took place over the Gulf of Tonkin. "My background is that of an expert in the observation of aircraft, principally military aircraft, but other aircraft would be in the area, in the daytime and at night time. I was also a qualified celestial navigator. So, I spent many days at sea. I spent an hour in the evening and an hour in the morning observing the heavens, the stars, the planets, and their relative positions. I'm a qualified observer in this respect.

"My initial encounter with a UFO occurred on a ship, when I was in the navy, in Vietnam, in 1968. My ship was following an aircraft carrier in the Gulf of Tonkin. We were a half a mile behind the carrier in case anyone should fall off, while they were engaged in airplane operations. It was after dark, when there were no planes flying. We were informed in advance if there were any planes or helicopters flying. We also were told if there were any planes returning from emergency missions, where they had to land in the carrier unannounced. We were the plane guard ship and were the first to be told. I was the duty officer. I was on the bridge, in control of the ship.

"Suddenly, out of the sky came three bright white lights. They swept down out of the sky and stopped off the left port side of the carrier. We had nothing on radar. The aircraft carrier went into a hard right turn. I unfortunately, had to navigate the ship, so we wouldn't have a collision.

"So, everyone ran off to the port side to watch the UFO, while I had to go to the starboard side to keep the ship following the carrier. I kept glancing over, as much as I could safely, to see what was happening. The lower light of the three went out, then the middle light went out, and then the top one went out. That was it. The carrier straightened out in formation and we continued with our operations.

MOST RECENT SIGHTING

"On July 12th of this year my wife and I had a sighting. We were returning home from a lecture we had attended in Yonkers. It was about 11:30 p.m. We were going north on the Taconic Parkway and were about five miles south at the Yorktown reservoir. There was a pattern of lights up ahead, which looked like an arc of bright lights. Last winter I had seen a pattern of red and white lights like that and I had thought it was an aircraft with really strange lights on it, and it just kept going. This time it was just up ahead and we were getting closer to it.

"Suddenly, the white lights went out. My wife said, 'Did you see that; it just disappeared?" I looked in the area where I had seen the lights and I could see dim red lights. They were clear but far away. I called her attention to the lights. She was able to see them also. We continued to move. I couldn't tell if it was moving or hovering. My wife wanted to pull off to the side of the road, but I told her to wait until we got closer.

"As we came over the hill, just near a reservoir, it was at a 75 to 30 degree angle off the highway from where we were. It was a little to the west of the highway. We went down the hill and across the bridge. It gave the impression, while we were moving, of a 747 coining in for a landing, with lots of funny lights on it. The bright lights were like what you would expect from landing lights, which weren't directed towards you but sort of away from you. You could see them. We went a half mile beyond the bridge, crossing the reservoir.

"It was now above us at about 75 degree angle from the horizon, a little to the left of the parkway. We pulled to the side of the road and got out of our vehicle. We were now in the woods, in a deserted area. There were no visible city lights. There was a full moon, just coming up over the horizon. It was a cloudless night, making it difficult to judge distances or sizes. But it gave the impression of being close and very large. It had 24 red lights."

A POSSIBLE PATTERN

Eugene Bower, in addition to being a careful observer himself, has tried to find a possible pattern for the sightings in upstate New York. "One night I started to collect papers with all the reports. I had my co-workers running around collecting everything they could find in print about the UFOs. I noticed that all the sightings during 1983 occurred from the first quarter until the full moon. There were only one or two sightings that didn't fall into this pattern."

After priming the audience for possible future encounters. Bower proceeded to tell about what got him interested in the subject. "My own sighting occurred on March 26th 1983. It was a day I'll never forget. It was a Saturday evening and we had just left a restaurant in Hyde Park. It was about 7:30. My companion and I decided to take a short cut. We took this very dark winding road. Suddenly, the young lady turned to me and said, 'My goodness. I don't know who lives over there, on top of that hill, but they have the brightest lights that I have ever seen. How could they have a spotlight that bright?' I turned to her and said, That's impossible, there's no one living there.'

"So, I quickly stopped my car. I left the lights on, as well as the radio. When I turned to look, I saw this thing, with four lights on one side and four lights on the other side and one in the middle. It was moving slowly over the power lines. At that point, I shut my car off and stepped out in front of my car. It was above my

UFOS DEJA VU

right shoulder. I could see it moving along the high tension wire. I thought it was going to crash into the wires. It was about the size of a 747. It was just flowing slowly.

"Then it stopped and raised itself. Then it moved until it was almost in front of us and turned. It seemed to look at us and dipped down. The girl got hysterical and said, 'Let's get the hell out of here.' I said, 'No, this is probably going to be the last time in our lives we are ever going to see something like this.' I wish that I could get to see it again. In fact, when I drive on the parkway, I have one eye on the road and one eye in the sky.

"After it faced us, the four lights on one side, the four lights on the other side, and the one light in the middle started to dim. It almost went out but it didn't. It made a crackling noise, which I can't describe. I'm not an expert but it seemed like it might have been getting power from the power lines, because of its position. At that moment the lights changed from white to red, then purple, then blue, then green, and then orange. This continued for about a minute. It was as if it were trying to say 'We are here. You are looking at us.'

"Then, the tights suddenly dimmed again and turned white. It turned on its side and I could see what looked like... it had the bones of a fish. We were about 200 yards away, as it turned, it went almost north, but then swung around to the back side.

"I ran a little bit down the road, because I wanted to see if it had a dome on it. There wasn't any dome. There was no sound. As it turned and started to go back it went almost east. I jumped back in the car and nervously fumbled with the keys to get the car started.

"When I got the car started I said to my friend, "We're going to make the first right hand turn and try to catch up with it." However, the trees were high and it could have been only fifty feet away and we wouldn't have seen it. We went up and down all of the roads looking for it. Then, I decided to get on the parkway, where there is an overview. We stayed there until 12:30 but we never sat it again."

And while there can be a lull in the activity that can last months, maybe a couple of years, almost everyone is certain that the UFOs will return at any moment to the New England Triangle, but the question remains — what is their intention?

UFOS DEJA VU

According to MUFON Cape Code has been inundated with UFO sightings in recent times, including craft that hovered near a famous lighthouse. Inspired art © by Carol Ann Rodriguez.

UFOS DEJA VU

The abduction experience of Betty and Barney Hill has been recognized by the state of Massachusetts with this iconic road marker.

More Hudson Valley **UFOs**

Including western Connecticut, northern New Jersey, and beyond...

Linda Zimmermann

Above: Linda Zimmerman's book gives the details of the massive Hudson Valley/New England UFO wave.

Left: Drum roll for first UFO whistleblower. Massachusetts Bay Colony co-founder, Governor John Winthrop, recorded observation of UFOs in the sky over Boston.

UFOS DEJA VU

Crowds gathered on the highway to crane their necks at the hovering triangle-shaped UFO overhead. Hudson Valley UFOs were mainly triangular.

UFOS DEJA VU

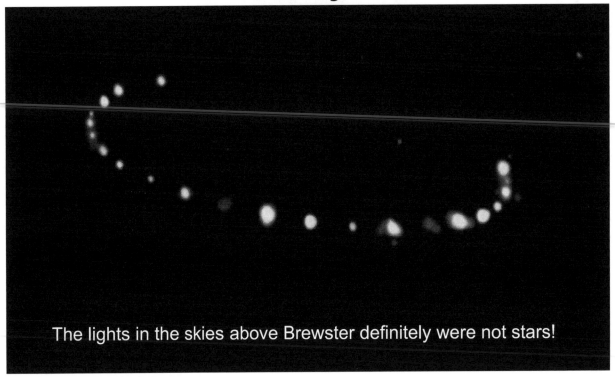

The lights in the skies above Brewster definitely were not stars!

Science teacher Philip Imbrogno and Dr. J. Allen Hynek question witness to close hand UFO

UFOS DEJA VU

SOME CALL IT THE "THING!"
By Timothy Green Beckley

At one point this relatively quiet British community near the Stonehenge monuments was the most active vortex in the world. Thousands came in all kinds of weather to successfully catch a glimpse of all manner of strange objects in the sky. Not lights in this case, but what seemed to be solid craft that you could pick the shape and size of to your satisfaction.

It was in 1981 that I traveled to the UK at the request of my good friend, the late Earl of Clancarty, Brinsley Le Poer Trench, who had arranged for me to speak at the House of Lords in front of a special group organized to get to the bottom of the UFO mystery and to press for "full disclosure," long before the phrase was part of UFO terminology. The group, consisting of roughly one hundred members of both houses of Parliament, included Lord Hill Norton, the former Defense Minister who had taken a combative interest in the subject of our unidentified visitors. Norton and I had a brief chat about Nikola Tesla, whom the Retired Admiral of the British fleet held dear to his heart because of the possibility of alternative energy sources being sparked by otherworldly life forms.

I arrived in Warminster and had lunch at one of the best Indian restaurants I have ever set foot in – and, believe me, I have eaten in hundreds all over the world. Among those joining me was my buddy, Arthur Shuttlewood, who was the editor of the daily newspaper "The Warminster Journal" and whose book "UFO Prophecy" I had published in the States. The book brought to the attention of the American public the story of an ongoing UFO flap that was taking place only a hop, skip and a jump from Stonehenge. Accompanying Arthur Shuttlewood was his sky-watch companion, a retired RAF pilot named Bob Strong.

While we munched on an appetizer, Strong showed me several scrapbooks filled with literally dozens of photos of UFO craft of all shapes and sizes, from "railroad cars" to huge, bat-like objects. Unfortunately, Strong sadly admitted that many of his best pictures had gone "missing" because they had been borrowed by the curious who wanted to copy them but had never returned them as they promised to do. Thus was lost an important part of UFO history that can never be replaced, as well as essential evidence proving the Warminster mystery was not a hoax or based upon faulty eyesight or mushroom-induced hallucinations.

That night, I journeyed with my new found compatriots to Cradle Hill, a few miles from the center of town, where we gathered with a couple from Scotland who had come on their own after having read about the town's ethereal intruders. They did so without knowing that they would soon be joined by the two gentlemen who had literally put Warminster's "Thing" on the landscape and a traveling "thrill seeker" from across the pond.

THE NIGHT I COMMUNICATED WITH A UFO

At first we saw nothing unusual but were fascinated by several meteors streaking across our line of vision. All that was to change around 10 P.M., when we spotted something unusual fairly high up that seemed to just be hovering or loitering about. To what intended purpose, if any, we had no way of knowing. Cueing us that this was no twinkling star or planet, Shuttlewood went to the trunk of his car and retrieved his trusty, high-powered flashlight. He told us he had used this same heavy-duty torch upon numerous occasions to signal to what he assured us were his Space Brother friends. Arthur then pointed it at the object in the sky and flashed a beam of light several times in its direction. He next offered the flashlight to me, and I also flashed it at the object. None of us knew ship-to-shore or any kind of Morse code, so we just blinked the light on and off like we were playing a game of "close encounters."

The reaction was unexpected and tremendously positive. Every time we blinked at it, the UFO would appear to sort of swing back and forth, like a pendulum. It seemed to be looking down on us. Maybe it picked up our thoughts telepathically. I was told to keep a positive mind because that's what the ultra-terrestrials seemed to respond to the most.

Our skywatch went on for another 20 minutes or so, and then it started to rain and clouds obscured our view. When we saw nothing more, we retired for the evening.

Had I made contact with a UFO over Warminster? To this day, I still wonder. I will never know for sure, but it did seem as if the object was under intelligent control and was responding to our request to prove it was not just an ordinary object in the heavens. Another strange thing: We took several photos of Cradle Hill that night. When developed, one of them showed a strange phenomenon behind two of the witnesses – streaks or bolts of light which were not visible to the naked eye at the time and for which I am mostly certain there is no "legitimate" explanation. There were no street lamps or houses with porch lights nearby . . . only a vast open field where many a strange incident had taken place over a time frame of several decades and which still ignite a fire in the hearts and minds of many seeking the truth about the UFOs seen over this locale.

UFOS DEJA VU

Many had come as skeptics and had left the area as firm believers. One New York City radio deejay ended up having a multitude of encounters in Warminster over a period of days.

In the book, "UFOS: Keys To Inner Perfection," Bryce Bond's meeting with human looking ultra-terrestrials is described in detail and includes an illustration of the occupants provided by Bryce's close friend for many years, Marc Brinkerhoff.

Bryce reveals many incredible experiences that cannot go overlooked, such as the one he describes first hand for our edification.

"Warminster, at this time of night, even for a Saturday, was somewhat deserted. Only a few people ambled along the narrow streets. I felt that I had eyes on me all the way. It was a most unusual feeling. The small narrow streets, with high brick walls, sky overcast, and the town strangely quiet — maybe a prelude to what I was about to experience that night! After walking a short distance through narrow archways and flower-lined paths, I was amazed how lovely it smelled and how clean it was. On the hill, some of the group had their telescopes set up on tripods; others had binoculars and cameras ready.

"The thing that really struck me was how friendly everyone was. A good portion are very curious, another percentage are thrill-seekers, and the remainder, well, they just enjoyed being there with this warm, loving group of spiritual individuals, sharing stories and conversations . . . UFOs have been here for eons of years. History is filled with reports of strange glowing craft, of landings and contactees. But due to negative programming regarding these ancients, and fear of the unknown being magnified out of all proportion from mouth to mouth, it spread right into modern times. Only in these times we blame television and motion pictures for doing the damage: creating near mass-panic in the mind, showing these UFOs as hostile, coming down from the heavens to devour, murder and rape — and to gobble us up!

"This travesty of beings who are thousands or even millions of years ahead of us in technology, intelligence and spiritual intent. . . . What the masses do not understand, they fear. When they fear, they shoot and run. There are numerous reports that UFOs were shot at, out of panic — even by the military. Put yourselves in their shoes and think: What would you do if you went to their planet, or dimension, or universe, or another period in time? We drove over to Starr Hill, another ancient burial ground area. This sector is where the Romans built upon, with a few of the remnants still in evidence. The location was down in a valley, wheat fields all around and high hills. The sky started to clear, filled with thousands of beautiful stars and still no UFOs . . .

"It was getting awfully late and still I had not interviewed Arthur Shuttlewood. My voice was getting weaker, my head clogged up due to the cool dampness. I got his attention and we crawled into one of the nearby cars to keep warm while I

UFOS DEJA VU

interviewed him. He was telling me that, only a few nights ago, three large entities about eight feet in height were seen down in a little hollow, to which he pointed. While in their presence, people felt a great warmth exuded from them; they were engulfed by it and the scent of roses and violets were very strong. All of a sudden, while Arthur was speaking, his conversation went to a peculiar light that just appeared in the field in front of us. He was somewhat blasé about the whole thing.

BRYCE BOND ENTERS A UFO

"Arthur then said quietly: 'I'm very glad you are here tonight, Bryce. There in front of us is a UFO. Notice the triangle shape and colored lights going around? That is a very good sign.' It then started to lift off in a weird pattern — then just disappeared. I was flabbergasted! It was so close. While describing that one on tape for American listeners, another one popped up about 25 degrees along the horizon. This one was very brilliant white, while the other was a blaze of colored lights. The intensity increased as it raised itself very slowly, did a little dance in the sky, then took off and disappeared. But before it did, Arthur jumped from the car, borrowed a flashlight from someone and sent Morse code to it. It in turn sent back the same signal that Arthur flashed out. Then it flew off. This was the highlight of my British trip: a close sighting; yet I honestly felt spiritually close to the lights in the field."

Bryce began interviewing other watchers as to what they witnessed.

"I turned off the recorder and made a mad dash out into the field, went into a light trance state and asked higher intelligence to make contact again. Leaving the group, I made my way down to the hollow, where two nights before three entities were seen. Again I went into a light trance state for what seemed to be a few minutes only . . . I was awakened by my friends, who thought I had gone. I must have been there for about an hour. I truly do not know what transpired while I was there or in trance. I told my friends I would return shortly and they went back to where the others were standing. I then made my way slowly back to the parked cars and people. Now here is a strange thing: the wheat in the field next to me as I walked back up the dirt road was about waist high . . .

"I walked along the road very close to the fence. Suddenly I heard a noise — like something crushing the wheat down. There was no breeze blowing that night. I looked over. The moon had just come out, shining very brightly — and there, before my eyes, a large depression was being formed. The wheat was being crushed down in a counterclockwise position. It too was shaped like a triangle and measured about twenty feet from point to point. I stood there a few moments and experienced a tremendous tingling sensation — the same sweet smell — being engulfed by warm air. Not fully understanding what had happened, I walked up the road to get Arthur, my host.

"Speaking of the field, Arthur pointed out some landing impressions in the section fronting the farm barn: a circle about thirty feet in circumference, with another depression spotted, but this one in a long cigar shape. All the depressions, recently made and noticed, were in a counterclockwise fashion. After all this, I was very happy and thankful. My mission had been a success," wrote Bryce Bond, who was also research editor of "Beyond Reality," a leading American psychic, occult, metaphysical and UFO magazine. He says some nice but wholly undeserved things about me in his newsletter and describes his trip back to London from Warminster in Reg Bradbury's Kingdom Crusade van. "On the way we stopped at Stonehenge, on Salisbury Plain. With the moon casting its eerie light upon this ancient structure, it was quite awe-inspiring in early morning hours. This whole area is just teeming in legend and folklore, and you begin to wonder who constructed these megalithic giants . . . and for what purpose?"

2020 is the 55-56th anniversary of the appearance of "The Thing" over Warminster. The ongoing wave of sightings took on many aspects, as can be verified by reading the literature or watching one of the several YouTube documentaries on Warminster - https://www.youtube.com/watch?v=y1bzRRg5gb8

To some the "thing!" appeared as a giant cross in the sky as depicted by inspired artist Carol Ann Rodriguez.

UFOS DEJA VU

UFO Prophecy Update! — 50th Anniversary Edition was published by Inner Light and is available on Amazon.com. Join the "Warminster UFO Research Group" on Facebook.com for updates and new photos.

The Thing! — many experienced it for themselves. The UFOs over Starr and Cradle Hills came in all shapes and sizes. See Warminster UFO Research Group page on FaceBook.

UFOS DEJA VU

After apparently establishing contact with a UFO over Starr Hill, Tim Beckley took this photograph of two fellow observers, only to have a series of "bouncing" orbs appear behind them. This despite the fact that there were no house or street lights or cars anywhere nearby.

Journalist Arthur Shuttlewood strolls on path to Starr Hill where the bulk of the sightings of the "Thing!" were observed.

UFOS DEJA VU

ADJUNTAS – A PUERTO RICAN GATE TO THE UNKNOWN
By Scott Corrales

Forty years ago, bringing up the subject of UFOs in Puerto Rico would elicit two responses as to where the phenomenon could best be seen. One was at the Caribbean National Rainforest – better known as the enigmatic, jungle-draped peak of El Yunque – and the other was a far-less known one outside the island itself: the municipality of Adjuntas, nestled in the island's central mountain range.

The Caribbean island's coffee-growing region, high in the Cordillera Central, is far cooler than the Atlantic or Caribbean shores: temperatures in the low sixties are common, with 40 degree minimums recorded in 1980 and 2012. Rich in vegetation, the municipality would acquire a reputation as a UFO hotspot – a characteristic that endures to this day.

The UFO presence in the center of the island, however, went much farther back than the 1970s. During the wave of sightings that precipitated itself over Puerto Rico in the 1950s, reports from the neighboring town of Jayuya described the presence of "a flying saucer of considerable size" over the community on October 9, 1952 at 7:30 P.M. The light projected by the incredible object as it hovered over a sugar refinery was such that local schoolteacher, Aida Reyes – hoping for a better look – had to close her eyes at one point, saying that the brightness was comparable to that of looking at the sun through closed eyelids. She would later go on to say that the object's apparent size was larger than "a barrel lid" and oval shaped, moving silently across the skies.

Activity over Adjuntas proper began – according to journalistic sources – in October 1972 as part of the island's nearly constant wave of paranormal activity, which was not to be replicated until the 1987-1995 flap. Local newspaper El Nuevo Día mentioned the presence of unidentified flying objects whose manifestations had caused a commotion in Adjuntas and caused concern among the population. Police sources told reporters that the object was orange-yellow, would appear in the night skies and then vanish, only to appear again. It also followed a regular schedule, manifesting between seven and ten o'clock in the evening. The more skeptical – and now defunct - San Juan Star sided with the opinions of the local weather service, which dismissed the sightings as a weather balloon launched

UFOS DEJA VU

from Isla Verde International Airport every night at 7 P.M. The prevailing winds, according to the Weather Service, could have probably taken the balloon to the vicinity of Adjuntas, some fifty miles distant.

Luis Maldonado Trinidad, a police commander for the island's southern area (who would become a notorious police commissioner in the future), supposedly forwarded a report to his superiors, ascertaining that police personnel had in fact seen these unusual objects. His report describes them as "balloon-shaped yellow lights traveling at considerable speed while remaining noiseless, giving off a luminosity that lasts between fifty and sixty seconds. They travel from one side to another and vanish into the mountains."

Sebastián Robiou's Manifiesto OVNI gives us a more specific case from these saucer-ridden times. Around nine o'clock in the evening on October 10, 1972, Milton Kleber and his nephew were driving along one of the narrow roads to the north of Adjuntas on their way to Dos Bocas, a lake created in the 1940s as part of the island's hydroelectric network. As they drove, uncle and nephew saw an object, "rounded and flattened like an oyster" following a trajectory parallel to their vehicle. "To the left of the highway we saw the dark silhouette of a mountain against a sky filled with low, black nimbus-stratus clouds. Between the mountain and the highway, at an elevation of nearly a thousand feet, at a 60-degree angle with my car, we then saw the outline of an object moving in the sky along the horizon, at a speed equal to that of the car and matching it. It was pale orange, not shiny, but rather a matte color."

The experience does not end there. The passengers continued to observe the object and Kleber ultimately decided to pull over. "The time was now past nine o'clock in the evening. We were at kilometer marker 5.6 between two houses facing each other. The object also halted in the sky, keeping ahead of us but always to the left. Its orange light had turned bluish-green like that of a mercury lamppost. It then moved to our eleven o'clock position as the pale orange hue returned. It kept moving until it placed itself in front of us at a considerable distance."

It was then that Kleber decided to flash its headlights at the unknown object, which proceeded to disappear and reappear four times in succession. "What impressed us the most was not that it had responded to our signals, but the way in which it did so. It seemed to slowly absorb all of its own light until it became invisible, and then issue it gradually until it regained its former luminosity. Something like pulses of light in slow motion."

The travelers resumed their journey, reaching the town of Utuado, where they learned that police officers had been called out to investigate a series of phone calls from local residents, reporting the presence of strange things flying around in the heavens.

UFOS DEJA VU

THE IMPORTANCE OF OCTOBER

On October 12, 1972 another island newspaper, El Mundo, published photos allegedly taken in the Adjuntas, linking these images to the story of a teenage contactee in the city of Ponce who had entered into "telepathic communication" with otherworldly beings. On the next day, the newspaper ran a communiqué from the Arecibo Atmospheric Observatory, stating that the flying objects seen over Adjuntas and Ponce were little more than "gases emitted by the observatory into the atmosphere to clear the atmosphere when it is overloaded." One would think that the observatory's press officer delighted in posting such releases: in the 1990s, UFO sightings in the same area were explained away as "the observatory engaging in the de-orbiting of satellites."

But no amount of explaining away could allay local fears and excitement. Rigoberto Ramos, the mayor of Adjuntas, told the press plainly: "Had I not seen these things with my own eyes, I would have never believed in flying saucers." The politician had been driving to San Juan at night with his entourage when he was bemused by the sight of three discs resembling the full moon, varying in intensity as they crossed the heavens, vanishing behind the mountains. "There was no doubt in my mind," Ramos continued, "that I was witnessing something I had never seen in my life."

This affirmation of belief by the elected official caused all eyes to turn toward Adjuntas, and for thousands of curiosity seekers to jump into their cars and visit the mountain-girt community. Newspapers now devoted entire features to the sightings. A local factory worker declared that he too had seen the objects, describing them as oval-shaped, orange and silver colored, noiseless and flying from east to west, losing themselves amid the jagged mountains. An agricultural worker would describe the objects as looking like the sun, traveling at average speed before vanishing. A college student claimed having seen a "coffin-shaped" flying object from his grandmother's country home, flying at considerably altitude.

Researchers Noel Rigau and Sebastián Robiou visited the community during this critical period of the three-week long flap and managed to speak to dozens of witnesses, even securing an interview with Mayor Ramos, who in turn provided them with a list of reputable persons who had seen objects in the sky, ranging from the town priest to the school principal, including an array of students, merchants and farmers. Among them was Nino González, who reported an interesting CE-2: his television set and all of the electric appliances in his home went dark as a "flying saucer" flew overhead. Another local resident, William Serrano, would also claim seeing a light as he walked toward his house one evening, noticing how the lights within the structure browned-out as the strange light grew closer. When the light appeared to "head straight toward him", Serrano jumped into a nearby pile of brushwood to avoid it. Renowned ufologist Stanton Friedman ap-

UFOS DEJA VU

parently visited the area without obtaining any satisfactory evidence, following a presentation at the University of Puerto Rico in Rio Piedras.

The 1972 Adjuntas flap petered out toward the end of October, as the island was gripped by another kind of fever – political campaigning – and turned out in droves in November to elect Luis A. Ferré to the governor's office.

A refractory period of UFO activity would ensue for nearly eleven months, followed by some of the most startling cases ever recorded in Puerto Rico during the worldwide "year of the humanoids," of which we have written elsewhere.

In writing about the Adjuntas flap of 1972, the late Gordon Creighton of Flying Saucer Review (FSR CH 1973 No. 17) had the following to say:

"The number of people in the Adjuntas area who claim to have seen UFOs is remarkable. There, as all over the island, newspaper offices and TV stations and police stations were inundated with phone calls. [...] The crowds flocking to Adjuntas have caused the bars and amusement centers to flourish and everywhere great excitement has prevailed. Puerto Rico's last UFO wave (1968) was also over the central, southern and western areas. An interesting point that has emerged is that these rural areas of Puerto Rico, which appear to have been under inspection, are on, or close to, zones of high natural radioactivity in the soil, as has been confirmed by consultation of the U.S. Geological Survey's Map, National Gamma Aeroradioactivity of Puerto Rico. In all cases, the UFOs allegedly seen at close quarters are less than 16 ft. in diameter. An unusual, striking feature is that so many have been seen in close vicinity of reservoirs, dams, electric plants, high-voltage power lines and radio installations."

SIGHTINGS IN THE 1950S

In the tenth month of the year 1990, UFO activity returned to the mountains of central Puerto Rico, following the renewal of activity in Cabo Rojo (on the island's southwestern corner) in the late '80s. Residents of Adjuntas and Utuado began contacting newsrooms much in the same way they had done in 1972, reporting unusual craft settling on the surrounding mountains. One particular claim described a large, spherical object that disgorged smaller ones, all of which vanished in a westward direction.

Julio Víctor Ramírez, staff writer for the El Vocero newspaper, featured the story of watchman Heriberto Acosta in his column on UFOs. The watchman, a resident of Lajas, to south of Adjuntas and the mountain range, reported seeing a strange craft whose lower section was fully illuminated, and had the general configuration of "a turtle's shell." From his post at the Lajas municipal dump (we would do well to remember at this point the UFO phenomenon's affinity for such places), the watchman gazed incredulously at the huge object from a wooden bench on the ground.

"I looked to the sky and saw this thing above me, some sixty feet above, and I

317

got scared. There was nothing I could do; I couldn't take off running," he explained, adding that the unknown vehicle remained static for an unspecified period of time before suddenly ascending. "That thing looked like a giant turtle, and I saw that it looked whitish underneath, but it had no lights. Still, it cast light as though it were daytime."

Those citizens of Adjuntas who remembered the madness of the early '70s, were not exactly pleased to witness a resurgence of the phenomenon, especially when it seemed to bring with it other anomalies like the "snowfall" which turned out to be a spectacular hailstorm that dropped inch stones on the population for half an hour. But not even the atmospheric phenomena could keep the curiosity-seekers away, who clustered by the dozens, becoming witnesses to the strange objects that made free with the area's airspace. It was interesting that both spectators and unknown lights should be drawn to this spot: the location of the test pits for potential mining operations in central Puerto Rico, known to contain considerable quantities of copper and to a lesser extent, gold. It has been speculated that this mineral wealth has attracted nonhuman interest as well, or that it serves as a beacon for anomalous activity.

Elderly residents of Adjuntas' Barrio Pellejas alleged that bizarre vehicles and even stranger-looking "people" could occasionally be seen in the region containing the copper mines. In an interview, Mrs. Rafaela Hernández indicated: "My father would tell us that there was something strange there, a great mystery. That those people supposedly from another world had a base there, and that late at night, saucers would land and a great glow could be seen down there."

A THRILLING SKYWATCH

Recordings provided to INEXPLICATA by Willie Durand Urbina of the Puerto Rican Research Group (PRRG) attest to the excitement of the skywatches during the 1990-91 flaps. An unidentified UFO enthusiast can be heard saying: "Ladies and gentlemen, now we are seeing another object! Two minutes after an airliner fly-by, we are now watching a triangular object, heading out of Utuado, stopping over Pellejas, on its way to [the district of] Juan González. It is currently over Juan González. This is incredible. How is it that so many objects are being seen in the area? It isn't an airplane. It flies very slowly, and it's about to fly past the antennas [TV masts] that can be seen from here. This object is heading toward Jayuya. What we do not know is why they wait for airliners to go by before they come out."

The unidentified male voice asks a female companion: "Machi, what is it you're seeing?"

"Well, right now it looks round, through these binoculars," she replies, "and curtailing its speed."

In the following seconds of audiotape, a hurried description is offered as to how an airplane – not a commercial airliner – appears to be engaged in reconnoitering the unidentified object, which pulls away. "The UFO is above the plane, flashing its lights. The plane is trying to find it." A confusion of voices ensues, with another man saying that the unknown quantity "is heading off toward Cerro Maravilla", Puerto Rico's fourth highest peak (3,500 ft.).

A sighting captured forever in time.

A LETTER TO THE PRESIDENT

In October 1991, residents of the Jardines de Adjuntas urbanization began reporting UFOs flying over their community toward Barrio Guilarte. Noel González, a respected local civic leader, reported seeing a strange object of impressive proportions. The object projected a white beam of light and moved slowly and noiselessly.

At the height of the flap, it was said that some of the enormous steel plates employed to cover some of the copper test pits dug by Kennecott during decades past had been torn asunder by an unknown force, possibly beams emanating from UFOs. Photo and video evidence alleging the destruction of the enormous steel plates covering the test pits was circulated around this time. The Adjuntas events reached such intensify that Rigoberto Ramos, the town's mayor, felt the need to contact President George W. Bush to apprise him of the situation:

"Adjuntas is a little town in the Central Range of Puerto Rico, and at this moment, we are very intrigued by some unusual events that are affecting our daily lives. Some years ago, we noticed the presence of Unidentified Flying Objects (UFO) in our skies. At first, we did not give great importance to this matter, but lately these things have appeared again and our citizens are distressed over this. Many, many persons have witnessed the presence of these objects in our surrounding space (evidence of these apparitions is included). Our purpose in writing you is to ask for your help to clarify what is really happening by ordering an investigation so that the people in our community can keep calm."

Functionaries attached to the Adjuntas town hall did not say whether the White House had replied to the mayor's missive.

[A political side note: The Hon. Rigoberto Ramos Aquino was reelected to the mayor's office in 1976, having lost to his opponent in the 1972 elections. Undaunted, he ran again in 1976, winning another four-year term, losing the reelection during the 1980 campaign. He threw his hat into the race again in 1989, winning a third term until 1996. He died in May 2015, mourned by the leadership of the island's New Progressive Party.]

In November 1992, journalist Julio Victor Ramirez reported on a fascinating – and frightening – incident in the mountain community: uncommonly large luminous objects were reportedly descending on a hill known as El Gigante in the precise sector of the municipality where the mining test pits were located. The witnesses to the event included members of the local police department, who

UFOS DEJA VU

would subsequently retell the experience to radio personality Edwin Plaza: "That thing had a set of lights beneath it, and a white ray of light issued from its bottom, lighting up the hill." Despite its tremendous size, the vehicle made no sound whatsoever.

According to the law enforcement agents, the massive luminous object had appeared in the early hours of the evening, bathing the slopes of El Gigante in white light. The unknown object reportedly had lights underneath it and the source of illumination came from a single beam projected against the hillside. What made this sighting interesting – as if its magnitude were not sufficient to make it important – is that the object's beam was apparently seeking a particular location: an agricultural school on the slopes of El Gigante that looked into "improved cattle ranching techniques." With what we know about the presence of the UFO phenomenon in the worldwide epidemic of cattle mutilations, is it unreasonable to suspect a connection in this case?

SUGGESTED READING

INEXPLICATEBLOGSPOT.COM

UFO HOSTILITIES

ALIEN BLOOD LUST

AREA 51 WARNING KEEP OUTINEXPLICATEBLOGSPOT.COM

SCREWED BY THE ALIENS

CHUPACABRA AND OTHER MYSTERIES

Author Scott Corrales in a pensive mood.

UFOS DEJA VU

Above: Adjuntas Casa Alcaldia
Below: Mountainous area where many encounters occurred.

UFOS DEJA VU

The rugged and nearly unpopulated mountains around Adjuntas.

Below: Artist's rendition of the humanoid beings and their unusual craft.

UFOS DEJA VU

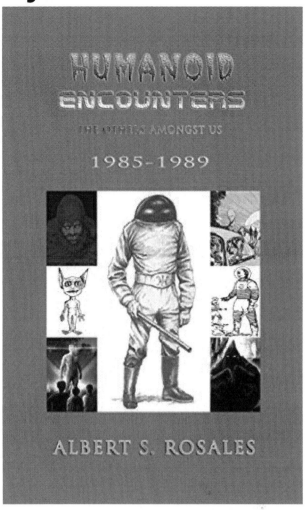

Albert Rosales' book detailing humanoid beings encountered in and near Adjuntas, Puerto Rico.

Three-term mayor Rigoberto Ramos, UFO witness

UFOS DEJA VU

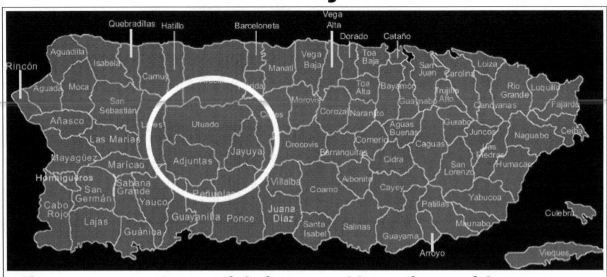

The municipalities of Adjuntas, Utuado and Jayuya

An artist's rendition of the humanoid beings observed in the area in and around Adjuntas, Puerto Rico. They appear to be "beaming down" — or up— to or from their craft in some sort of light column.

SECTION TEN
DEJA VU ONE LAST TIME

JUNE 24TH, 1947: Businessman pilot Kenneth Arnold, flying his CallAir 2, side number NC33355, near Mount Rainier, Washington, observed nine tailless cresent shaped aircraft flying between the mountains in a manner described as "...like saucers if you skipped them across the water." And, thus, was born the era of the "Flying Saucers."

UFOS DEJA VU

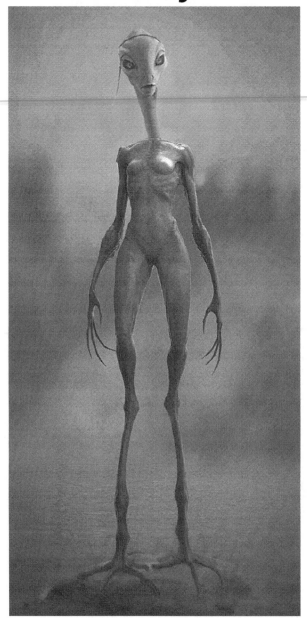

SECTION TEN

CHAPTER THIRTY-TWO
WHAT HAPPENED ON JUNE 28TH 1980?
By Albert S. Rosales

CHAPTER THIRTY-THREE
DEJA VU TIL THE END OF TIME
THE JUNE 24TH – KENNETH ARNOLD – ENIGMA
By Timothy Green Beckley

UFOS DEJA VU

SO WHAT HAPPENED ON JUNE 28TH 1980?
By Albert S. Rosales

PUBLISHER'S NOTE: We all have good days. We all have bad days. Days are days you might think. But that isn't necessarily true in the crazy – wild! — world of UFOlogy, especially in the arena of Deja Vu. Albert Rosales is a top notch researcher of humanoid cases. His files rival those of the Aerial Phenomena Research Organization. He gathers data from many sources. Enough to fill more than a dozen books. Albert was born in Cuba and migrated to the US in 1966. He witnessed several unusual incidents as a young man while living in Cuba, which continued throughout his life in the US. Albert became interested in unusual phenomena and UFOs at a young age, but soon directed his focus to the crux of the phenomena...the humanoids and otherworldly entities. He began collecting data on encounters from worldwide sources in the late 80's. His current database has over 20,000 entries, which is updated and corrected daily. One of his most popular works is "UFOs Over Florida: Humanoid and other Strange Encounters in the Sunshine State."

* * * * *

In the convoluted history of UFO/alien sightings and encounters, there are so-called flaps (also called waves) and window areas, places in which apparently at certain times, days and locations there is an increased number of strange encounters, with UFOs, humanoids, etc. Why does this happen? There are many theories, like how window areas are said to be thruways in which alien craft arrive at Earth from wherever they are from, and sometimes the concentration of cases at a certain location (sometimes even a country [example: France 1954]) is overwhelming and filled with high-strangeness.

These flaps or waves don't necessarily involve the same location, but sometimes occur on the same date and in widely separate locations on the globe. Case in point, June 28, 1980: On this date there were five very interesting encounters, spread out around the world - all of them of a very high-strangeness quality.

I would like to present them in order (time) of occurrence. The first occurred in the morning in Argentina, which did experience a wave throughout the year of 1980. It involved an encounter with a bizarre human-like entity and other assorted

UFOS DEJA VU

phenomena, perhaps connected to the main encounter. Here it is:

* * * * *

Location: Rafaela, Santa Fe, Argentina. Date: June 28, 1980. Time: morning. The night before (June 27th) members of the Carelli family were already in bed as a result of the intense cold temperatures. Then one of the family members, "Malvina," heard noises coming from outside her window. She woke up Mrs. Carelli and both heard what appeared to be "slow moving heavy footfalls" coming from the patio stairwell and later on the terrace area. Minutes later Malvina saw a strong light moving outside the window that shone right through the bedroom curtains.

Mrs. Carelli also observed the light. They did not go outside to investigate. On the morning of the 28th, Mrs. Carelli was outside hanging some laundry and noticed that it was very cold. Her pet dog "Esteban" accompanied her. As she worked she noticed, about 20 meters away, a woman that was accompanied by two children. The two children, described as blond and wearing very light clothing in relation to the weather, were playing around a huge cactus plant. The woman wore a tight-fitting brown colored outfit, again kind of light for the weather. She had long blond-platinum hair and large beautiful penetrating, blue-green eyes.

Mrs. Carelli approached the stranger and warned her about the dog, but the woman seemed unconcerned and said, "He will not bite us." The witness further describes the woman as very delicate and "beautiful", with a lean voluptuous body.

Mrs. Carelli then asked the woman if she was cold, but she replied in the negative.Moments later, Mrs. Carelli, a little perplexed, walked back into the kitchen; from that vantage point she no longer could see the strange woman and the children.

An hour later Malvina went outside to check and saw the woman, this time breastfeeding a baby, and sitting next to the cactus while the two other children remained playing nearby. There was no obvious mode of transportation around (vehicles, etc). Malvina approached and also spoke to the woman, who looked at her with the penetrating eyes in a somewhat disdainful stare.

The dog walked around the strangers, barking but did not dare approach. (The dog was later seen in the kitchen area and apparently disappeared later and was never seen in the area again).

On that same morning at a nearby sports field witnesses found strange tracks on the ground resembling an isosceles triangle. Also, mysteriously, the water level from the sports pool had inexplicably and substantially decreased. Source: Eduardo Adrian Ficaretti FAECE-ONIFE

* * * * *

The second case occurred a few hours later in Russia, which again was experiencing a sort of mini-flap at the time of the bizarre encounter by a witness who had previously witnessed other bizarre events: Location: Tslachtchevo, Moscow re-

UFOS DEJA VU

gion, Russia. Date: June 28, 1980. Time: 1400. Vladimir Pavlovich Bogachev had, back in August of 1979, encountered a bizarre invisible barrier in a field near his home. Earlier on this date he had been sitting outside with his young daughter when both heard a loud helicopter like sound over the nearby trees. They did not see anything, however. Later on, as he was walking along a field near the village, he again encountered the invisible barrier and became tired and his throat became dry. He sat down in a clearing and suddenly heard a loud pitched voice in Russian, "Do you want to see girls"! it exclaimed.

Looking up he saw a very tall man-like figure wearing what appeared to be a heavy monk like outfit. The figure smiled at Bogachev and he stood up. At this point the humanoid began to disappear before his eyes, first becoming vertical stripes and then completely vanishing. Moments later the humanoid reappeared again, this time wearing what appeared to be a tight-fitting shiny "spacesuit," green in color. The suit covered him completely except for his head.

The stranger was heavy set with European features. He had a pointy chin and high cheekbones and what appeared to be scraggly beard. He repeated his original questions about the girls and disappeared again. In his place the witness could now see a meadow-like panorama, blue and white in color, resembling a rectangular screen.

On the low cut grass there were a group of lovely girls, wearing white thin fabric like clothing and headgear. They had beautiful long black hair and were barefooted. Their skin was snow white in color and all had wonderful proportionate and voluptuous bodies. They somehow resembled Japanese girls but with European features, about 17-years of age.

One of them spoke in a melodious but incomprehensible language. The witness then heard a hissing sound and the scene vanished. A fog came over him and he seemed to lose consciousness. He soon found himself entering an object with white linoleum like walls. On a wall hung a poster, similar to a hand painted drawing; on it there was a picture of a five pointed star. To the right of the room he saw large green curtains that led to a larger room. Behind the curtains he encountered two apparatus like devices and a table in the center and what appeared to be comfortable sofas on each side.

On the ceiling there was a round fixture with a bright light. In this room he saw a very tall man, about 2 meters in height, with short black hair on each side and bald on top, wearing a narrow trouser with a belt and a short sleeve striped shirt, brown in color. On the right stood another man wearing gray clothing that completely ignored the witness. The tall alien spoke to Bogachev, asking him what he feared most. The witness immediately answered, "A nuclear war". The aliens promised to become involved in the affairs of men and to prevent a catastrophe.

After promising to contact the local scientific community Bogachev was deposited back in the same field. Apparently, he did not see the aliens again.

Source: Boris Chourinov, Cuadernos De Ufologia Vol. 15 # 2 1993

The third and perhaps the most odd encounter occurred on the coast of Puerto Rico, which was apparently what I call a "permanent abduction."

* * * * *

Maldonado and his aircraft.

Location. Mona Channel, Caribbean Ocean near Puerto Rico. Date: June 28, 1980. Time: 2003. At 1810 Jose Antonio Maldonado Torres and his friend, Jose Pagan Santos, took off from Las Americas International Airport in Santo Domingo, Dominican Republic in an Ercoupe aircraft marked N3808H. The Ercoupe was owned by Santo's father Jose Pagan Jimenez, an Aero Police officer in Puerto Rico. They were bound for home in Puerto Rico.

At 2003 the Las Mesas radar site and several aircraft picked up radio transmissions from N3808H: "Mayday, Mayday, Ercoupe ocho cero, eight zero, zero, Hotel. We can see a strange object in our course. We are lost, Mayday, Mayday."

An Iberia Airlines flight IB-976 en route from Santo Domingo to Spain responded to the Mayday and received a reply:"Ah we are going from Santo Domingo to ah San Juan International but we found ah a weird object in our course that made us change course about three different times. We got it right in front of us now at one o'clock, our heading is zero seven zero degrees…our altitude one thousand six hundred, at zero seven zero degrees…our VORs got lost off frequency…"

Iberia Flight IB-976 then relayed a message from San Juan Center asking N3808H to turn on their transponder. N3808H replied that the Ercoupe was not equipped with a transponder. At 2006 Iberia IB-976 asked for their call sign and estimated position and received this reply: "Right now we are supposed to be at about thirty five miles from the coast of Puerto Rico but we have something weird in front of us that makes us lose course all the time. I changed our course a second (unintelligible). Our present heading right now is three hundred. We are right again in the same stuff sir."

They were not heard from again. At 2012 the Atlantic Fleet Weapons Range verified the last radar position of N3808H as thirty five miles west of Puerto Rico. A search that included Santos's father was then mounted which centered on this last radar position. It was discontinued after two days when no trace had been found. No trace was ever found.

Source: http://www.ufocasebook.com/n3808hpuertorico.html

Comments: It seems to have been a permanent abduction oddly similar to the Valentich disappearance of 1978 in Australia and the cabin cruiser "Witchcraft" off the coast of Miami in 1967. As an interesting aside, according to researcher Scott Corrales, months later Jose Antonio's mother reportedly saw him in a vision one afternoon. He was wearing a metallic green uniform with black metal boots, and told her that he had joined the "extraterrestrials" in their mission and was quite safe and very happy with his new life.

* * * * *

The fourth encounter, an apparent abduction, was again from Argentina and the same province of Santa Fe. Details are brief but interesting. Location: Rosario, Santa Fe, Argentina. Date: June 28, 1980. Time: 2330. The witness was walking home alone and as he passed a large empty lot he noticed an intense light descending from the sky. As he looked at the light he felt attracted to it and was lifted up inside. He then found himself sitting on a large armchair surrounded by four short humanoids in yellow outfits and helmets. The beings stared at him silently. He was found 15 hours later next to a wall of the local cemetery.

Source: UFONS # 133

* * * * *

The last case was another abduction report from Arizona, an area filled with high-strangeness events: Location: White Tank Mountains, Arizona. Date: June 28, 1980. Time: 2330. Three witnesses had driven to an isolated mountain area when they all apparently drifted off to sleep. The main witness later awoke in a trance to see three small men wearing helmets and diving suits surrounding their vehicle. All three witnesses then exited the car and were floated over a nearby canal and towards a silvery dome shaped craft on the ground.

As they neared the craft they could see another little man standing on a rim that protruded around the bottom of the object. The little man pointed something resembling a flashlight at the main witness, then a rectangular opening became visible and all three witnesses entered the object. Inside there was a room that appeared to be the main control room filled with buttons and display screens. Five little men staffed this area.

The main witness was then taken into another room filled with bright blue lights. Inside this room she met a very tall man whose face was covered with a veil with symbols on it. She was then directed to sit on an L-shaped device after removing her clothing. The tall man communicated with her and apparently performed several tests and examinations on her. The little men later returned her and the other two witnesses to the car.

Source: William Hamilton, Alien Magic.

So what happened on that day indeed? A tare between other dimensions? A busy day for the Cosmic Trickster? We are only left to wonder. June 28, 1980 should not be forgotten (at least within Ufological circles).

But it is nothing compared to the strangest coincidental date of them all – JUNE 24th!

UFOS DEJA VU

UFOS DEJA VU

UFOS DEJA VU

Albert Rosales is the author of numerous books on humanoid encounters in what I like to call the "generational series." Albert's website is crammed with information. Visit it! http://ufoinfo.com

UFOS DEJA VU

DEJA VU TIL THE END
THE JUNE 24TH – KENNETH ARNOLD – ENIGMA
By Timothy Green Beckley

I have a secret . . . I know something you don't know.

I have made what I believe is a shocking, but monumental, discovery involving the connection between UFOs and life after death, and thus the portals of time itself.

It is almost too strange and too bizarre to be revealed, but I must do so, as disclosing "the truth" is an essential part of the UFO "game."

Kenneth Arnold, for all intents and purposes, is regarded as the literal father of modern day UFOlogy. On June 24, 1947, he sighted a formation of rapidly moving objects that appeared to be skipping over "water" as they moved in-between the cloud-covered peaks of Mount Rainier in Washington State.

Arnold was flying rather low, looking for a military plane that was said to have gone down somewhere between Tacoma and Yakima (been there, done that) , when his eye caught something glittering in the brilliant sunlight.

"I saw the flashes were coming from a series of objects that were traveling incredibly fast," he would say soon after the experience. "They were silvery and shiny and seemed to be shaped like a pie plate. What startled me most at this point was that I could not find any tails on them."

While Arnold passed away in 1984, his sighting over the years has been elevated to what might be considered a "saintly" status in the eyes of most UFO believers. In fact, the anniversary of Arnold's sighting is celebrated annually around the world as UFO Day. If you go to "Mr. UFO's Secret Files" on YouTube and do a search for Shanelle Schanz, you will come up with an episode of a podcast I hosted that ties in with this story. "The Strange Paranormal Saga of Kenneth Arnold's Granddaughter" consists of an intriguing interview with someone very closely related to perhaps the most important historical figure in the world of unidentified flying discs.

Shanelle Schanz has an elaborate tattoo gracing her upper back, from shoulder to shoulder. It is a fabulous rendering, showing the date of her grandfather's sighting along with an illustration of one of the objects he observed which has become etched in the memory of so many of us who take the subject of UFOs seriously. On the program, Shanelle dropped a number of bombshells, at least to the average listener not familiar with anything but Arnold's initial observation. Truth is, Arnold had a total of eight sightings of unidentified craft in his a career, plus a run-in with the dreaded Men In Black and a possible attempt by a government agent to silence him "for good."

But this missive is about Kenneth Arnold, UFOs and matters of a supernatural nature, primarily about life after death and the connection between UFOs and the hereafter. It is the ultimate gateway, the portal that you will experience last. It is the way we began this book with some stuffing revelations about our journey past the ultimate gateway as revealed by Brent Raynes and Diane Tessman. And now it will be Deja Vu no more!

Now this is where the subject at hand gets downright spooky. Creepy. Eerie. Use any assortment of adjectives you might like.

Because, you see, this missive is about Kenneth Arnold, UFOs and matters of a supernatural nature, primary life after death and the connection between UFOs and the hereafter.

I have discovered over the course of many years that so-called coincidences and synchronicities play a major role in the UFO phenomenon. I mean, at one time or another, we have all thought about an individual we haven't seen in ages and five minutes later the telephone will ring and it will be that individual on the other end of the line. Hey, that's interesting, but it doesn't prove much of anything.

The synchronicities in my life are much more dramatic and seem to be under the direct command of someone or "something." Some incidents that have happened to me are way beyond the pale. I have a 400 page book, "The Matrix Control System of Philip K. Dick and the Paranormal Synchronicities of Timothy Green Beckley" if you want to delve into the bizarre nature of coincidences and the paranormal. I've talked about the topic on Coast to Coast and elsewhere so it might not be earth shaking news to some of you – for others it will blow your MF mind.

You see, what most people would refer to as a simple "coincidence" is more than just a random chance occurrence. And somewhere out there someone – or something – is directing the "show" and trying to draw attention to some other reality. Often these "coincidences" seem to be evidence of life after death.

THE JUNE 24TH ENIGMA

Now here the kicker, the punch line, the really strange part of this and when it finally starts to come together . . .

As we have ascertained, June 24th is a very important day in the history of modern day UFOlogy, the date when Ken Arnold had his sighting.

UFOS DEJA VU

But what isn't generally realized is that several – no make that numerous – UFO investigators and authors have passed away on this historic day, as if the "Grim Reaper" – or "someone – had hand-selected this date to usher the spirits of these famous UFOlogists from this world to the next.

I call it the June 24th Enigma!

Here is a list of those associated with the field who passed away on June 24th as best as we can determine.

FRANK SCULLY – The author of "Behind The Flying Saucers," the first best-selling book on flying saucers (circa 1950), a book that introduced the topic of crashed space ships and dead space beings to the public. Scully was a columnist for Variety and a humorist. In October and November 1949, Scully published two columns in the highly respected show business trade publication claiming that extraterrestrial beings were recovered from a downed flying saucer near Aztec, New Mexico, based on what Scully said was reported to him by a scientist involved. He died at the age of 72, on that "memorial" date of June 24th, 1964.

FRANK EDWARDS – After WWII, the Mutual Broadcasting System hired Edwards to host a nationwide news and opinion program sponsored by the American Federation of Labor. Edwards' program was a success and became popular nationally.

Edwards began mentioning UFOs on his radio program, and wrote several books on the subject. His "Flying Saucers, Serious Business" is thought to have sold more copies in hardback than any other UFO book in publishing history.

He was dismissed from the radio program in 1954 for reasons that remain unclear. His interest in UFOs was believed to be a factor.

Edwards died the night before he was supposed to speak before the National UFO Conference, held in New York by "Saucer News" publisher Jim Moseley. It was SRO audience, but Edwards would miss it because it was June 24th once again!

LYLE STUART – An even more odd twist came about when Lyle Stuart – who was the publisher of Frank Edwards' books – also died on June 24, though not the same year as his author's passing. Stuart was thought of as being pretty much of a maverick publisher, putting out controversial tomes that no one else would publish, such as an exposé of Scientology. Stuart first gained national notoriety by taking on the powerful newspaper columnist Walter Winchell in a series of scathing magazine articles, collected in book form in 1953. After serving with the United States Merchant Marines and the Air Transport Command in World War II, he worked for William Randolph Hearst's International News Service, Variety, Music Business and RTW Scout. I used to sell some of his books related to the paraphysical at my Occult Center.

JACKIE GLEASON – Actor, comedian, musician, pool hustler and paranormalist, the "Honeymooner's" star must have known what he was talking about when he threatened to send his TV sitcom wife "to the moon, to the moon, Alice," because he might have already been there – or close enough to see the lunar surface at least. As a private collector, Gleason had one of the largest depository libraries on psychic phenomena and UFOs. When I was 16, he sent me a letter and a check to purchase a copy of one of my earliest books, "UFOs Around The World." That book, as well as thousands of others from his depository on the occult, now rests in the hands of the University of Miami. Gleason lived in a circular, flying saucer-style house and was a guest from time to time on the Long John Nebel "Party Line" show, which was the first all night radio program devoted to the strange and unexplained.

Gleason also is said to have gone with golfing buddy Richard Nixon to catch a glimpse of an alien in a deep freeze stored in a secluded hanger at Homestead AFB, Florida. I remember there was supposed to be a UFO update on ABC's Nightline on June 24, 1987, to commemorate the fortieth anniversary of Arnold's encounter, but that segment of the show was presented in abbreviated form in order to add a last minute commentary about Jackie Gleason's passing that very same day.

CONGRESSMAN MARIO BIAGGI – (October 26, 1917 – June 24, 2015) was a U.S. Representative from New York (served from 1969 to 1988) and former New York City police officer. He was elected as a Democrat from The Bronx in New York City. In 1987 and 1988, he was convicted in two separate corruption trials, and he resigned from Congress in 1988. He was one of the few politicians who took a serious interest in UFOs and didn't seem fazed that he might be subject to ridicule because of his beliefs. Report magazine.

DR. JAMES MARTIN – This esteemed Oxford scientist, futurist and computer systems scientist was found floating face down in the Bermuda Triangle off his private island. In 2000 he authored a well-received book "After the Internet: Alien Intelligence." He died on June 24, 2013, at age 79.

WILLY LEY – Willy Ley was a German-American science writer and space advocate who helped popularize rocketry and spaceflight in both Germany and the United States. The crater Ley on the far side of the Moon is named in his honor. He was one of the first respected modern scientists who took a crack at answering the question of what is a flying saucer. He was one of the first, if not the first, person to say that 85% of all sightings could be identified, but what about the remaining percentage? Ley died at the age of 62 on June 24, 1969, in his home in Jackson Heights, Queens, New York.

ROBERT CHARROUX – Charroux was a pioneer of the theory of ancient astronauts, publishing at least six nonfiction works in this genre in the last decade of his life, including "One Hundred Thousand Years of Man's Unknown History" (1963, 1970); "Forgotten Worlds" (1973); "Masters of the World" (1974); "The Gods Unknown" (1974) and "Legacy of the Gods" (1974). His death on June 24,1978, came as a surprise to everyone.

Of course we do not know the actual number of UFO researchers who have passed into – we hope – heavenly realms on June 24 as there is no international obituary for researchers of unexplained aerial phenomena. There could be many more. Your guess is as good as mine, but if you know of others who fit into this category please drop me a line at mrufo8@hotmail.com

THE ARNOLD HYPOTHESIS

Well, if it seems we were diverted in our drive to get to the bottom of this June 24th synchronicity in which various researchers have proceeded into the great abyss, it is time we get back onto our interdimensional Route 66 and proceed to our cloud covered destination, heaven.

For a long time I have tried to explain to people that Arnold did NOT believe that the objects he saw on several occasions were physical ships from one or more planets. He saw them as some sort of heavenly "conveyer belt" (my term) to the other side. He didn't quite make this a secret, but I suppose no one really pressed him on his innermost thoughts concerning UFOs, and for that matter Arnold had a great mistrust of the press.

UFO theorist Mike Clelland has posted on his "UFO Experience" blog some fascinating comments about Arnold, some of which I have heard before and can confirm. Arnold's thoughts have been published here and there, but never in one place as Clelland has done.

"Arnold's experiences went well beyond that initial event in 1947," Mike notes. "Arnold went on to see a number of other UFOs throughout his life; he reported that UFOs could read his mind; he and his family saw floating orbs in their home; he claimed his phone was tapped; he was threatened by the military to keep quiet about what he knew and he was fascinated with synchronicities. He came to see these events as happening to him for a reason and he eventually saw the whole thing as a spiritual experience. Arnold also came to believe that the UFO phenomenon might represent some kind of connection between the living and the dead. All this and a pet owl on his ranch!

"Arnold had another sighting that involved a cluster of about 25 small craft. He later had yet another sighting over California in 1952. He was in his plane and flew above two distinct craft. One was 'as solid as a Chevrolet,' the other was semi-transparent, and he could look down on it from above and see the pine trees on the ground through the center of the object. He sensed these objects had the ability to change their density, seeing them as living organisms."

Here's what Kenneth Arnold said as far back as 1967: "The impression I had after observing these strange objects a second time was that they were something alive rather than machines – a living organism of some type that apparently has the ability to change its density similar to [jelly] fish that are found in our oceans without losing their apparent identity."

Arnold, notes Mike Clelland, had some bold ideas about UFOs in an era of nuts-and-bolts thinking. He wrote about his beliefs in the November 1962 issue of Ray Palmer's Flying Saucers From Other Worlds magazine: "After some 14 years of extensive research, it is my conclusion that the so-called unidentified flying objects that have been seen in our atmosphere are not spaceships from another planet at all, but are groups and masses of living organisms that are as much a part of our atmosphere and space as the life we find in the depths of the oceans."

Arnold had:

++ Fascination with synchronicities

++ Belief that UFOs were somehow connected to the dead

So, in short, Arnold believed that the UFOs he sighted were some sort of semi-living organisms, vessels that were responsible for taking the souls of the recently departed over to the other side.

Now, if this hypothesis is true, wouldn't it stand to reason that they would be personally transporting the souls who respected them and spoke so lovingly about them in life. The spirits of Frank Edwards, Frank Scully, Jackie Gleason . . . well, we have named some of them a bit earlier.

Now Arnold is not the only individual to take into consideration the possibility that these craft could be some form of "mechanical angel" of sorts.

The popular AboveTopSecret.com web site has posted some comments in their chat room from those attracted to this concept.

In December, 2012 "Mandroids" followed up on this theory that had been generating some remarks:

"Could UFO's be vehicles that enable the passed on to visit our dimension? One military remote viewer spoke of seeing his deceased father aboard a UFO. Some describe the afterlife as another dimension or universe where we move to. Quantum immortality is also an interesting theory. I think this. Most departed know it and are offered brief trips to see us or our 'realm' to say hello. Perhaps transcendental beings offer this? A many worlds tourist trip."

It didn't take long for fellow chat roomer poster "Bluesma" to chime right in about an incidents involving his deceased mother.

"I must admit, I have wondered about the link between our departed and UFOs. The reason being that when I saw one in plain daylight, and quite close, hovering and then maneuvering over me while my car stalled with everyone else's on the road, I had the sensation of being communicated with through my head. It was very strange – as if I was being communicated with telepathically. What struck me as strange is that I had the impression my mother was one of the energies doing so, and emanating from the craft . . . but my mother had died a couple of years before!

"I couldn't make heads or tails of how she (or her soul, or whatever) could have anything to do with a craft like I saw. It was your typical flying saucer shape, made of some sort of metallic material, and made impossible maneuvers. But it seemed very real and physical – not like lights in the air, or something vague like that, with which I would have found it easier to come to conclusions of a 'spiritual' nature.

"There were 'thought packets,' which at that time I wasn't very adept at unraveling into linear form yet, but could grasp only the general gist, which, in the case of her communications, was some sort of explanation (as I was immediately asking questions) of this being a state she 'returned' to, or otherwise went on to . . . though whether this state is actually physical (like if she was some sort of physical three dimensional entity up there) was not clear at all."

Native Shamans worldwide have long professed that they have communicated with the other side. Many in a psychedelic state have drawn objects and beings that closely resemble UFOs and what we call "aliens." Harvard University's Dr. John Mack spent time in the Amazon conversing with the region's inhabitants and was surprised to learn that there seems to be an umbilical cord between this world and the next. UFOs could be a way to pass through this portal from one universe or dimension to the next.

Can we get to Heaven in a UFO? Is the afterlife a lot closer than we think?

We hope you've discovered some meaningful secrets within the pages of this work.

It has been a long, strange journey we have found ourselves on as we end our investigation – if only for now.

Hopefully, I will see you soon across the gates of time.

Comments And Personal Experiences Welcome

Mrufo8@hotmail.com

Mr. UFO's Secret Files channel on YouTube.com

https://www.youtube.com/user/MRUFO1100

Website

www.ConspiracyJournal.com

www.TeslasSecretLab.com

Podcast – Exploring the Bizarre – KCORradio.com – Every Thursday 10 PM eastern, 7 PM Pacific.

UFOS DEJA VU

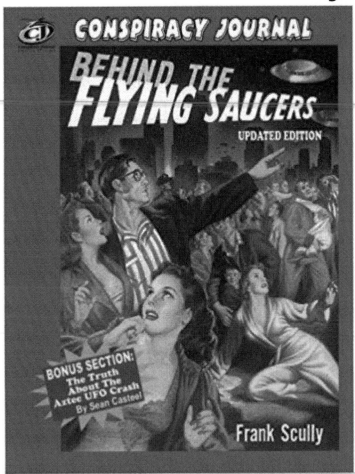

Frank Scully wrote about the crash of a UFO in the New Mexico desert.
We have our own special — Expanded! — edition of "Behind The Flying Saucers" available on Amazon.
Scully also passed on June 24th. Too eerie for words.

Jackie Gleason had a massive UFO collection. He passed on the anniversary of Arnold's sighting. A coincidence? Hardly!

UFOS DEJA VU

Thousands around the world celebrate the anniversary of Kenneth Arnold's landmark sighting. Most UFO buffs do not know that death has taken over a dozen researchers on June 24th.

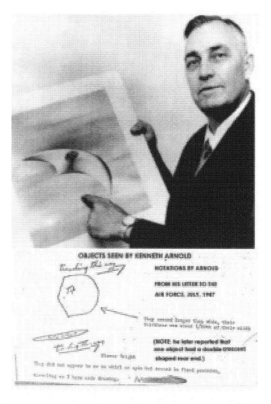

As a tribute to her grandfather, Ken's granddaughter has had a tattoo of the saucers he observed over Mount Rainier inked on her shoulder.

Kenneth Arnold. Right time, right place, right man.

Timothy Green Beckley's **ISSUE #50**   **US $3.00**

Conspiracy Journal
bizarre bazaar

Incorporating Inner Light / UFO Review

Promoting Free Speech and Individuality
Opposing The System, Censorship, Death & Taxes

IN THIS ISSUE: Strange, Amazing, Rare and Exotic Items From Worldwide Sources

A LONG AND WINDING ROAD—JOIN US ON OUR EXPLORATION OF THE UNKNOWN REALMS

"Nothing to see here. Move along please."

That's what you are usually told at the scene of an accident. But it's no accident you're receiving this publication in the mail, and there is every reason to sit back, relax and take a sip of something cold while the summer months roll along.

No one can accuse us of being a "one trick pony." We can almost guarantee that there is something within these pages for everyone. Some of you like your aliens (but do they like you?), while others are hiding in the brush waiting for Bigfoot or some other cryptid to wander by (be sure to have your camera ready). And let's not forgot those occultists, New Agers and spell crafters who love our more spiritually orientated material. Oh and hey, there are probably even a few items tucked away for those who are into more conventional Bible studies. We had a good size hit on our hands recently with Rev. Barry Downing's **"Biblical UFO Revelations,"** and Sean Casteel's **"Search For The Pale Prophet."** (They are, of course, still available — see pg 16 of our last issue, far right, second row of books for a two book special).

But hey let's jump to the present now, as we have some wonderful new titles to toot Gabriel's Horn about (don't you just cherish the way I phrase things?).

We all love our pets. They are sometimes much closer to us than our so-called friends and family. Maria D' Andrea tells us in her latest volume **"Witchcraft, The Occult And How To Select A Familiar,"** the best way to discover the importance familiars can play in our occult and witchcraft studies to bring about a world of

GEF
THE
TALKING MONGOOSE

positive benefits to the spells we cast and the rituals we perform.

Familiars can help with healing, making relationships go well, and in obtaining prosperity. They can also help the shaman do magic and dispense advice.

Witches have used them as spies thanks to their shape shifting abilities.

Now here's a pet I know you don't own. **"Gef The Talking Mongoose"** was the 8th Wonder of the World.

He lived on the Isle of Man, near the rugged coastline, could speak the Queen's English (with a few nasty words tossed in), had the abil-

Continued >

HENCHMEN
Tim Beckley, Publisher/Editor
Associates: Tim Swartz; Sean Casteel; Carol Ann Rodriguez
Layout, Graphics & Typesetting: William Kern (Adman)
CONSPIRACY JOURNAL GLOBAL COMMUNICATIONS
Payment For All Merchandise To Timothy Beckley
Box 753, New Brunswick, NJ 08903
MRUFO8@hotmail.com - PayPal Orders Preferred
All Other Methods Accepted—253-602-3407
FREE VIDEOS on our YouTube Channel
"Mr. UFOs Secret Files"
Tim Beckley

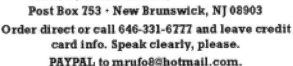

347

353

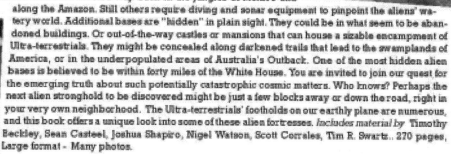

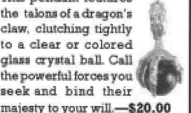

360

Made in the USA
Middletown, DE
15 December 2019